# SAE EDGE™
## RESEARCH REPORT

# Unsettled Topics Concerning Airworthiness Cybersecurity Regulation

## Aharon David
### *AFUZION-InfoSec*

---

**EDGE DEVELOPMENT TEAM**

Dror Ben-David, *Neural Networks R&D Lab (NRDL) at Matrix*

Daniel DiMase, *Aerocyonics Inc.*

Vance Hilderman, *AFUZION-InfoSec*

Angeliki Karakoliou, *EASA*

Kirsten M. Koepsel, JD, *Independent Aviation Cybersecurity & Counterfeit Parts Expert*

Patrick Mana, *EUROCONTROL*

Daniel Nebenzahl, *Resilience Cyber Security*

Antonio Nogueras, *EUROCONTROL*

Chris Roberts, *Hillbilly Hit Squad*

Cyrille Rosay, *EASA*

Peter Skaves, *FAA Advisor*

Chris Sundberg, *Woodward, Inc.*

**SAE INTERNATIONAL®**

Warrendale, Pennsylvania, USA

## About the Publisher

SAE International® is a global association of more than 128,000 engineers and related technical experts in the aerospace, automotive, and commercial vehicle industries. Our core competencies are life-long learning and voluntary consensus standards development. Visit sae.org

## SAE EDGE™ Research Report Disclaimer

SAE EDGE™ Research Reports focus on topics that are dynamic, in which knowledge is incomplete, and which have yet to be standardized. They represent the collective wisdom of a group of experts and serve as a practical guide to the reader in understanding unsettled subject matter. They are not meant to provide a recommended practice or protocol. The experts have assembled as a community of practitioners to contribute and collectivize their thoughts and points of view; these are not the positions of the institutions or businesses with which they are affiliated, nor is one contributor's perspective advanced over other contributors. SAE EDGE™ Research Reports are the property of SAE International and SAE alone is responsible for their content.

## About This Publication

SAE EDGE™ Research Reports provide state-of-the-art and state-of-industry examinations of the most significant topics in mobility engineering. SAE EDGE™ contributors are experts from research, academia, and industry who have come together to explore and define the most critical advancements, challenges, and future direction in areas such as vehicle automation, unmanned aircraft, cybersecurity, advanced propulsion, advanced manufacturing, Internet of Things, and connectivity.

## Related Resources

**SAE MOBILUS® Cybersecurity Knowledge Hub**
https://saemobilus.sae.org/cybersecurity/

## SAE Team

Frank Menchaca, Chief Growth Officer
Michael Thompson, Director, Standards, Information and Research Publications
Monica Nogueira, Content Acquisition Director
Beth Ellen Dibeler, Product Manager
William Kucinski, Managing Technical Editor

**EPR2020013**
**ISSN 2640-3536**
**e-ISSN 2640-3544**
**ISBN 978-1-4686-0189-3**

**To purchase bulk quantities, please contact:** SAE Customer Service

E-mail:    CustomerService@sae.org
Phone:    877-606-7323 (*inside USA and Canada*)
          +1-724-776-4970 (*outside USA*)
Fax:      +1-724-776-0790

https://www.sae.org/publications/edge-research-reports

# About the Editor

**Aharon David** is a co-founder and partner of AFUZION-InfoSec, a global services, consulting, and training company specializing in aviation cybersecurity certification. He is also a speaker and trainer on aviation cybersecurity certification for organizations such as SAE International, AIAA, IEEE, Aerospace-Tech-Week, and others. He is a member of all US and European standard-making committees for aviation safety-critical electronic systems, cybersecurity, artificial intelligence (AI), and unmanned aircraft systems (UAS), including SAE's S-18, G-32, and G-34; RTCA's SC-216 and 228; European Organisation for Civil Aviation Equipment's (EUROCAE) WG-72, WG-105, and WG-114; and others.

For the last decade, he has been an advisor to Israeli government authorities, such as the Civilian Air Authority of Israel (CAAI), on subject matters such as UAS, avionics software, cybersecurity, and more.

As an aerospace engineer and an Information Systems/Technology Management MBA with nearly four decades of hands-on engineering management and executive experience with the Israeli Air Force (IAF), Israeli Ministry of Defense (MoD), CAAI, and others—he held such positions as the commander of the IAF's Avionics & Control Software Centre (ACSC) and head of the Israeli Missile Defense Organization's (IMDO) System Engineering & Interoperability Department-among others.

He combines perspectives from technology, business, management of large organizations, civilian and defense, and passenger and UAS aviation development and certification.

# contents

# Unsettled Topics Concerning Airworthiness Cybersecurity Regulation

## Abstract

The certification process of the Boeing 787, starting in 2005, marked a watershed for airworthiness regulation. The "Dreamliner," the first true "flying data center," could no longer be certified for airworthiness ignoring "sabotage," like the classic safety regulation for commercial passenger aircraft. Its extensive application of data networks, including enhanced external digital communication, forced the Federal Aviation Administration (FAA), for the first time, to set "Special Conditions" for cybersecurity.

In the 15 years that ensued, airworthiness regulation followed suit, and all key rule-, regulation-, and standard-making organizations weighed in to establish a new airworthiness cybersecurity superset of legislation, regulation, and standardization. The resulting International Civil Aviation Organization (ICAO) resolutions, US and European Union (EU) legislations, FAA and European Aviation Safety Agency (EASA) regulations, and the DO-326/ED-202 set of standards are already the de-facto, and soon becoming the official, standards for legislation, regulation, and best practices, with the FAA already mandating it to a constantly growing extent for a few years now—and EASA adopting the set in its entirety in July 2020. This emerging superset of documents is now carefully studied by all relevant actors—including industry, regulators, and academia—as the aviation ecosystem moves forward with DO-326/ED-202 set training, gap analysis, and even with certification itself.

This report suggests a deeper analysis of these sets of regulatory documents and their effects on the aviation sector as they gradually become the law of the land, starting with their expected effects on the aviation ecosystem, the issues they pose to supply chains, and the challenges they present to the airworthiness certification process itself. Then, this report examines the major DO-326/ED-202 set gaps, inherent dilemmas, and methodological uncertainties. For each such unsettled domain, six aspects are reviewed. Finally, practical solution-seeking processes are proposed, and some specific potential frameworks and solutions are pointed out whenever applicable. It is the intention of this report that these insights and observations would assist regulators, applicants, and standard makers through, at least, the 2020s with accommodating this new regulation and start adjusting it to emerging realities.

NOTE: SAE EDGE™ Research Reports are intended to identify and illuminate key issues in emerging, but still unsettled, technologies of interest to the mobility industry. The goal of SAE EDGE™ Research Reports is to stimulate discussion and work in the hope of promoting and speeding resolution of identified issues. SAE EDGE™ Research Reports are not intended to resolve the challenges they identify or close any topic to further scrutiny.

**AHARON DAVID**

*Chief WHO (White Hat Officer), AFUZION-InfoSec*

### Edge Development Team

Dror Ben-David, *Neural Networks R&D Lab (NRDL) at Matrix*
Daniel DiMase, *Aerocyonics Inc.*
Vance Hilderman, *AFUZION-InfoSec*
Angeliki Karakoliou, *EASA*
Kirsten M. Koepsel, JD, *Independent Aviation Cybersecurity & Counterfeit Parts Expert*
Patrick Mana, *EUROCONTROL*
Daniel Nebenzahl, *Resilience Cyber Security*
Antonio Nogueras, *EUROCONTROL*
Chris Roberts, *Hillbilly Hit Squad*
Cyrille Rosay, *EASA*
Peter Skaves, *FAA Advisor*
Chris Sundberg, *Woodward, Inc.*

ISSN 2640-3536

# Introduction

Modern passenger aircraft are referred to as "flying data centers" [1]—and rightly so: from essential flight control and engine control interconnected systems and crew "glass cockpits" and portable electronic flight bags (EFBs) (Figure 1), through massive external communication systems such as the classic Aircraft Communications Addressing and Reporting System (ACARS) (Figure 2) and the modern Automatic Dependent Surveillance-Broadcast (ADS-B) (Figure 3), to in-flight entertainment and communication (IFEC) systems (Figure 4) and even passenger-carried personal electronic devices (PEDs), these aircraft are buzzing with digital communication.

As helpful as this trend might be in preventing critical errors, reducing aircrew workload, streamlining operations and maintenance, and improving passengers' flight experience, it also poses an imminent danger: cyber-threats, ranging anywhere from a negligible nuisance to a full-scale catastrophe or the shutdown of entire aviation segments for substantial periods of time [2, 3, 4].

Of the two innovative aircraft types to be introduced in the first decade of the twenty-first century, the Airbus A380 demonstrated a modern, fully networked cockpit (Figure 5) utilizing the then-new integrated modular avionics (IMA) concept, while the Boeing 787 "Dreamliner" (Figure 6),

launching its certification process in 2005, presented a fully networked aircraft, thus making it potentially vulnerable to cyber-threats. The Federal Aviation Administration's (FAA) reaction was to impose on the B787 "Special Conditions" (finalized in 2007) for the "Protection of Airplane Systems and Data Networks from Unauthorized External Access" [5]. This was the first of about 20 different aircraft types and type modifications to be assigned Special Conditions of this sort in the next decade [6], until, in 2014, the FAA issued a Policy Statement [7] for the "Establishment of Special Conditions for Aircraft Systems Information Security Protection" to put some order in the house, and updated it in 2017 [8] to at least define some rules for the ad hoc Special Conditions.

However, the B787 made the development of proper regulation imminent, and indeed, in 2006/2007, both the European Organisation for Civil Aviation Equipment (EUROCAE), by establishing WG-72 [9], and RTCA, by establishing SC-216 [10], followed through and started the development of such guidance, methods, and considerations-for the European Aviation Safety Agency (EASA) and FAA to later adopt and make "Acceptable Means of Compliance" toward certification of cybersecurity aspects. It took 12 years and an additional interim ad hoc committee [11] to carry out this monumental task, in a few rounds of documents and updates for some of them, until, finally, in 2018, EUROCAE and RTCA finalized a complete (but still not perfect, and still

**FIGURE 2.** **ACARS Graphic Display within a passenger airplane's cockpit.**

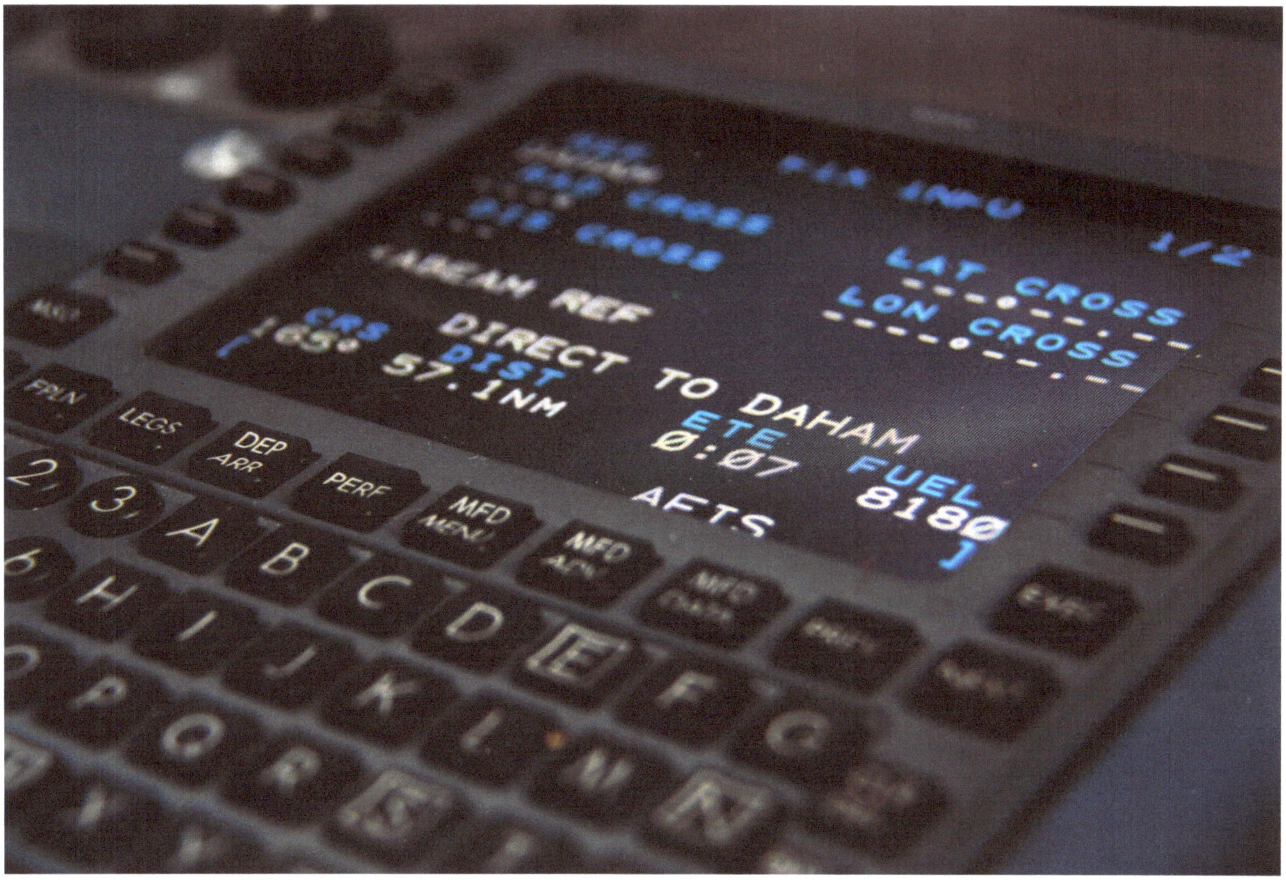


schusterbauer.com/Shutterstock.com

**FIGURE 3.** **ADS-B, FAA NextGen Implementation.**

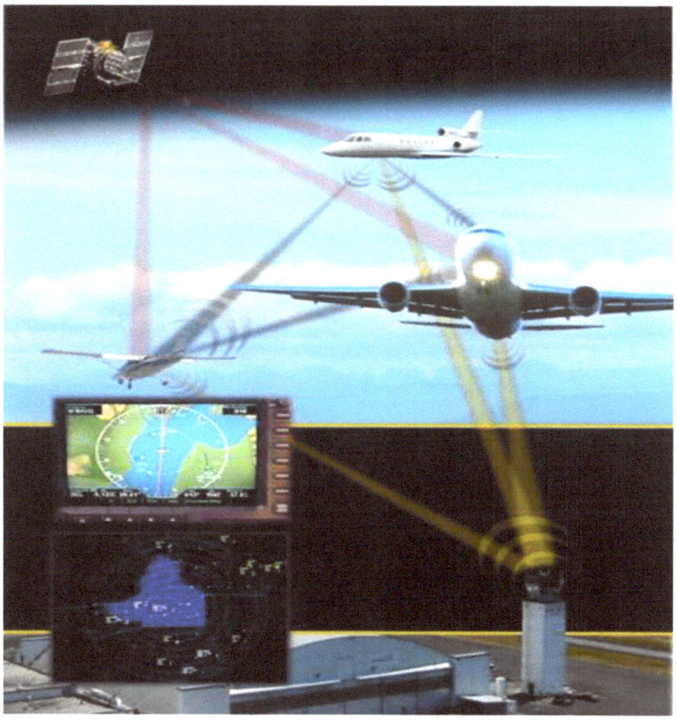

Reprinted from FAA

**FIGURE 4.** Typical in-flight entertainment and communication (IFEC) system.

**FIGURE 5.** Airbus A380 Cockpit.

**FIGURE 6.** Boeing B787 Cockpit.

Sorbis/Shutterstock.com

following on with updates) "DO-326/ED-202 set" (Figure 7) for airworthiness cybersecurity, including

- RTCA/EUROCAE: "DO-326A/ED-202A-Airworthiness Security Process Specification," 2014 [12]

- RTCA/EUROCAE: "DO-356A/ED-203A-Airworthiness Security Methods & Considerations," 2018 [13]

- RTCA/EUROCAE: "DO-355/ED-204-Information Security Guidance for Continuing Airworthiness," 2014 [14]

- EUROCAE: "ED-201-Aeronautical Information System Security (AISS) Framework Guidance," 2015 [15], to which RTCA intend to join with the next update

In Europe alone, EUROCAE introduced "ED-205-Process Standard for Security Certification/Declaration of Air Traffic Management/Air Navigation Services (ATM/ANS) Ground Systems" [16], in 2019, but since this document does not directly relate to airworthiness, it will not be further discussed in this report as part of the "set." EUROCAE also published two auxiliary documents for information only: ER-013 (Glossary) in 2015 [17] and ER-017 (InfoSec activity mapping) in 2018 [18]. A few updates to the set are already scheduled by EUROCAE and RTCA for 2020-2021, the most notable being a thorough makeover of ED-201, including a DO equivalent and a new DO/ED for ISEM (Information Security Event Management), which would basically be a spin-off of DO-355/ED-204. Beyond that, EUROCAE and RTCA can be expected to issue more updates and enhancements into the 2020s and beyond.

As EUROCAE and RTCA are not regulators, the FAA and EASA—as the leading airworthiness authorities—had to step in to make the DO-326/ED-202 set an acceptable means of compliance within amendments to existing regulation or new regulatory mandates.

However, the two organizations chose different paths. The FAA gradually issued "Advisory Circulars" (ACs) since 2015 that already directly or indirectly related to DO-326/ED-202 set for Aircraft Networks [19], Data Links [20], and EFBs [21], and proceeds with further ACs and with amendments to 14CFR, expected during 2020-2021. EASA chose a "one and done" approach with its RMT.0648 [22] producing NPA 2019-01 [23] to put all required regulation and even legislation in place, which was approved in July 2020 [24].

One potential complicating factor as of 2020 can become the United Kingdom's (UK's) withdrawal from the European Union (EU) ("Brexit"), as the UK is far more than a mainstream member as far as aviation is concerned: some of WG-72's most prominent members who were instrumental in the DO-326/ED-202 set development were UK representatives, and it is still unclear how Brexit will affect EASA and EUROCAE.

**FIGURE 7.** The DO-326/ED-202 set documents interrelations.

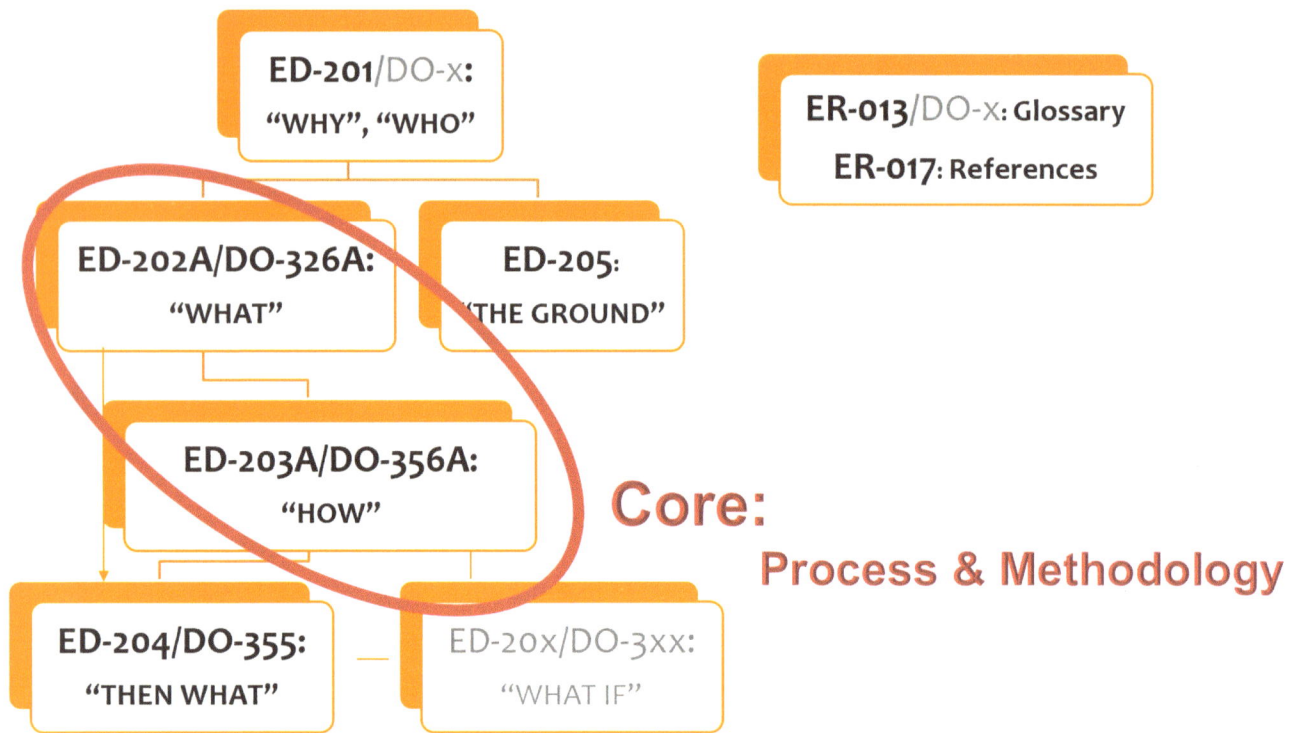

However, early indications from the UK's Civil Aviation Authority (CAA) show that, on one hand, very much like the FAA, the CAA is gradually introducing regulatory vehicles [25, 26, 27] to guide applicants but is poised to adhere to the DO-326/ED-202 set as the core instrument of such regulation, so this report assumes that the prominent CAA's regulatory activity will be very close in nature to either the FAA or EASA—or some combination of both approaches—at least in principle.

This report regards the DO-326/ED-202 set in its current version as accepted by the FAA and EASA for all practical purposes but will not refer to any updates of the set that were already made public as "unsettled."

To wrap-up the regulatory envelope, not only do other airworthiness regulators in the world adopt similar, if not identical measures [28], but global authorities weighed in as well: the International Civil Aviation Organization (ICAO), mandating essential steps through resolution A40-10 "Cyber-security in Civil Aviation" in its recent 40th Assembly [29], and more broadly International Air Transport Association (IATA), as well as other prominent organizations. This report will regard ICAO as the "global regulator," and since it is also recognized by IATA as such [30], no deep discussions of IATA will follow.

## State of the Industry

As of 2020, the most popular member of the DO-326/ED-202 set is still DO-355/ED-204 for continued airworthiness

InfoSec, as it is relatively simple, modular, does not require revolutionary organizational transformations, and is still relatively close to existing information technology (IT) cybersecurity standards, such as the ISO27000 family of standards [31]. Quite a few operators already use it voluntarily, even before it is mandated across the board.

The core documents of the DO-326/ED-202 set, DO-326A/ED-202A and DO-356A/ED-203A, are regarded as imminent across the entire aerospace industry, including the defense sector (to various degrees of acceptance). As of now, some certification processes using the DO-326/ED-202 set are underway, as prudent applicants already use the DO-326/ED-202 set even for Special Conditions cases, so as to gain experience and establish a better certification base for their next projects. Naturally, regulators and applicants alike now need to face quite a few unexpected theoretical and practical aspects of this new regulation with their recent undertakings.

## Unsettled Domains Concerning Airworthiness Cybersecurity Regulation

In order to touch the most critical unsettled topics concerning airworthiness cybersecurity regulation while remaining within a reasonable scope, a variety of experts, representing a wide array of vantage points, have been consulted with, and a few of these were gracious enough to become contributors for the report. They hail from air and ground, industry and academia, safety and security; they are defenders and ethical hackers, applicants and regulators, rule makers and standards makers, women and men—individuals from Asia, Europe, and the US. This report is the joint creation of all of these aspects and qualities.

Due to the magnitude of this field, and in order to keep this report within a reasonable size, it is confined to only airworthiness cybersecurity regulatory efforts taking place in the US and Europe, mainly the DO-326/ED-202 set, with minimal necessary references to other regions and aspects of aviation-at-large (which might be dealt with in future, follow-on reports).

This SAE EDGE™ Research Report considers the following six areas to be unsettled domains in airworthiness cybersecurity regulation, as marked by apparent concerns, challenges, gaps, dilemmas, and/or uncertainties:

- Aviation ecosystem and beyond ("ecosystem of ecosystems") concerns
- OEM supply chain issues
- Regulatory compliance challenges
- Current standards/guidance/best practices gaps
- Current standards/guidance/best practices inherent dilemmas
- Current standards/guidance/best practices methodology uncertainties

The following sections discuss the respective topics, challenges, and potential solutions for each of these areas in more detail. The goal of this SAE EDGE™ Research Report is to provide some perspective on where the regulation is heading in the essential field of securing airworthiness of large transport aircraft from Intentional Unauthorized Electronic Interference (IUEI) and suggest some recommendations for action by the community of practitioners.

# Aviation Ecosystem and Beyond: "Ecosystem of Ecosystems" Concerns

Airworthiness is, by far, the most crucial and sensitive subsector of aviation regulations, especially when safety is at stake—specifically, the safety of large commercial passenger aircraft. Nonetheless, airworthiness cannot, and does not, stand by itself—it is indeed a crucial part of the vast aviation ecosystem, which is, by itself, part of a global ecosystem of ecosystems. Various international organizations take part in developing and maintaining the aviation regulatory framework, which presents these major unsettled aspects:

- Slow aviation-at-large regulatory progress
- Cross-regulatory boundaries, cyber-attacks, and security
- Lacking mandated cooperation among stakeholders
- Insufficient regulatory progress outside the commercial-passenger-aviation sector
- Lacking regulatory harmonization with non-aviation ecosystems
- International mistrust/distrust: "Which side are you on?"

## Slow Aviation-at-Large Regulatory Progress

While regulation and standardization for airworthiness cybersecurity, rightly conceived to be the most important issue for aviation cybersecurity, is progressing at a reasonable pace and scheduled to be in place very soon, at least in the US and Europe—the rest of the aviation ecosystem is fractured territory, with multiple actors, not all of whom are synchronized with airworthiness needs. This aviation ecosystem landscape, as far as cybersecurity is concerned, has only recently made some hesitant moves toward ATM and ANS cybersecurity certification (in Europe only—the FAA goes it alone), while airport cybersecurity is just an afterthought of general airport security, not to mention domains that are even further away from the aircraft—physically and conceptually—such as flight booking.

Some of the progress made is under the scope of grand ATM modernization programs, such as NextGen (Figure 8) in the US, Single European Sky ATM Research (SESAR) (Figure 9) in Europe, and others—but these are mostly not coordinated with mainstream airworthiness cybersecurity rulemaking. At EASA, there are at least signs that the aviation ecosystem will be addressed into the 2020s, with RMT.0720 [32] being assigned the ambitious task of handling cybersecurity for all aspects of aviation cybersecurity, except for airworthiness (which is already handled by RMT.0648 [33]). The main concern with the current unbalanced regulatory state is that airworthiness might need, for the foreseeable future, to carry a substantial portion of the cybersecurity burden for the entire aviation ecosystem, which may turn out to be too heavy to handle. Obvious "weak links" of external aircraft communications, such as ADS-B and ACARS, have already been used as attack vectors and paths throughout the 2010s, demonstrating the potential implications for airworthiness safety [34].

Required steps to alleviate this concern could be along the lines of EASA's RMT.0720 (i.e., regulating, step by step, as many domains as practical of the aviation ecosystem with the end goal being a grand Part-AISS). Even so, a few issues would still require attention (e.g., setting the boundaries of each such domain, handshaking across domains, and even defining the outer boundaries of the aviation ecosystem for cybersecurity)

**FIGURE 8.**   **NextGen concept.**

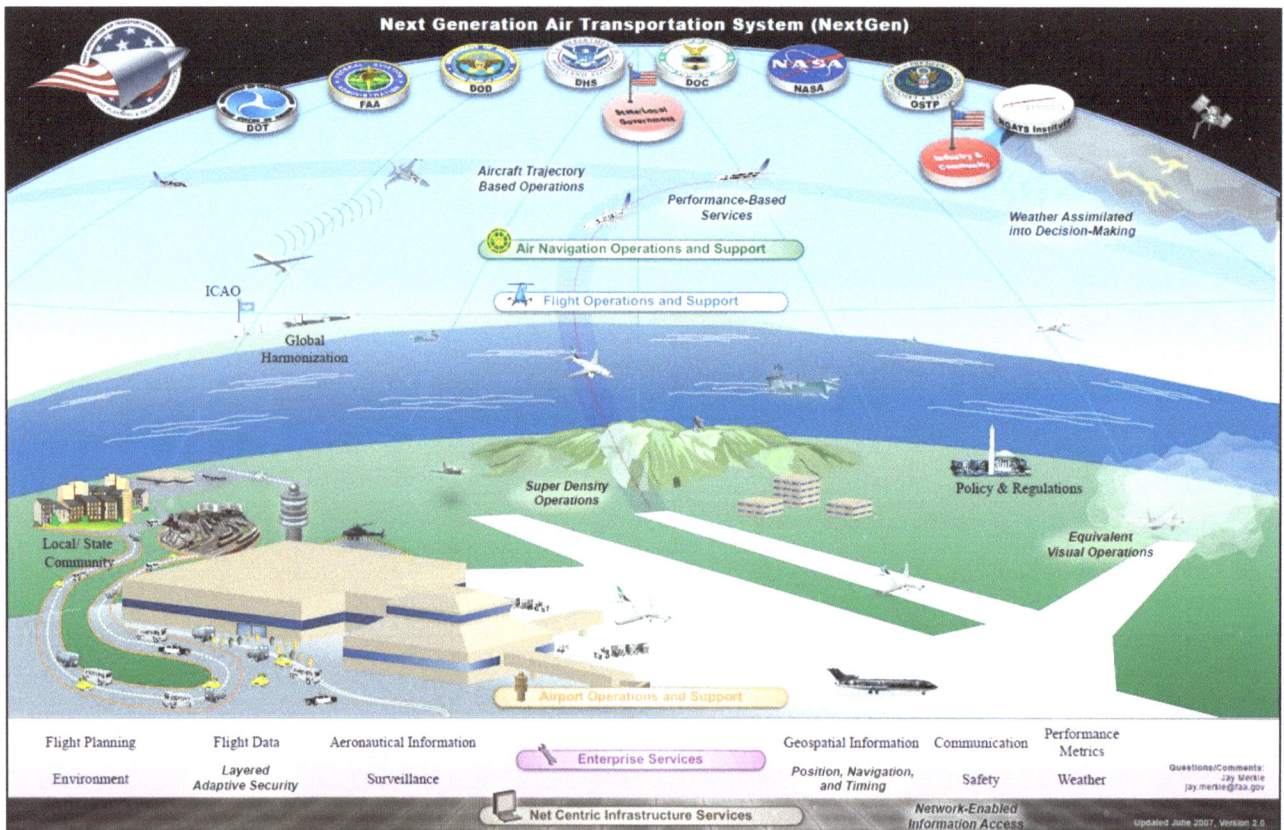

as cyber-threats tend to migrate across domains and ecosystems, both physically and virtually. The most practical approach would be to treat the aviation ecosystem inside-out (i.e., harden airworthiness first) proceed to the next in line (ATM, ANS, airports), and then advance to other subsectors as practical. This regulatory process should be stretched as far as possible throughout the aviation ecosystem, so as to put the widest buffer between aircraft and attackers.

It is advisable that, within this large process, its proper scalability and level of integration among the various aviation subdomains be defined.

# Cross-Regulatory Boundaries, Cyber-Attacks, and Security

Having mentioned aviation domain borders and handshaking across such borders—cyber-threats, as just stated, migrate across domains and even entire ecosystems. This results in another concern related to the entire aviation ecosystem and beyond: the very real possibility (that is explicitly referred to in the DO-326/ED-202 set) that attack vectors may originate *outside* the aviation ecosystem altogether. With this, is the understanding that—with near-certainty—such attack

vectors would originate outside the airworthiness domain of the aviation ecosystem. Such an eventuality makes the entire aviation ecosystem, and most certainly the airworthiness domain, but one element of the entire attack path.

Consider the simplistic, almost naïve, example mentioned in the previous section: an attack path that crosses from an ATM system (ACARS) to an aircraft. Which regulation should be in effect: ATM regulation (e.g., EUROCONTROL—extending beyond the EU), airworthiness regulation (e.g., EASA—an EU organ), both? If both regulators should be responsible, then how to design the allocation of responsibilities so that, at any given time/aspect, only one regulator is the "prime regulator?" In case regulatory authorities need to operate within each other's jurisdictions, should there be a "regulators' arbitration," pre-regulated, or should it be settled ad hoc? And to what extent would the airworthiness mandates diffuse into the ground systems? Would we then look at a meltdown of boundaries along the path of attack? There are many questions, but no reasonable answers yet.

Such cross-domain attacks yield regulatory boundaries almost meaningless for attackers and pose an array of concerns regarding the effective and efficient handling of cross-domain cyber-threats, as even risk assessments may become extremely challenging when extended across

**FIGURE 9.** SESAR concept.

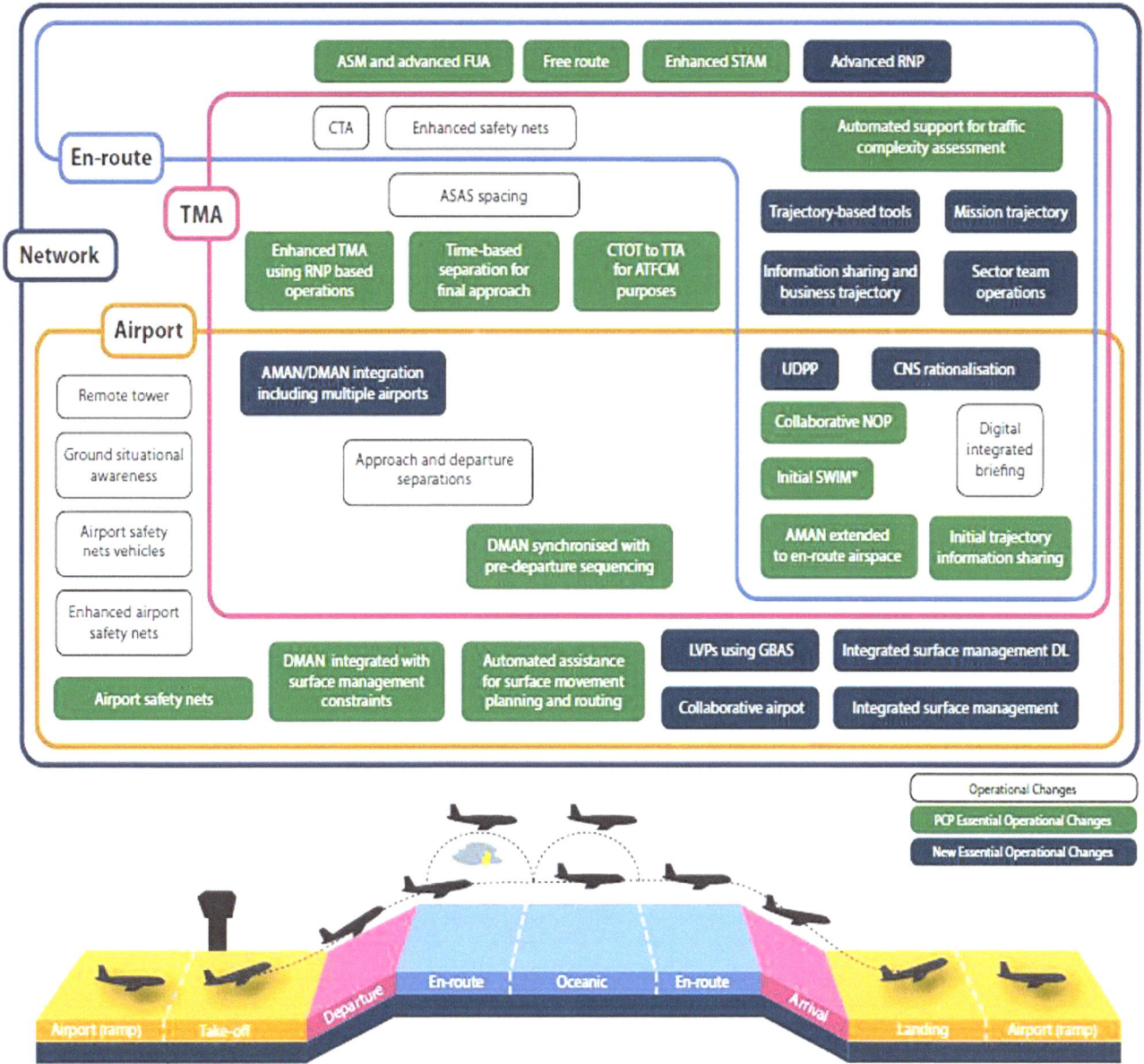

Initial SWIM* includes the following PCP Essential Operational Changes:
- common infrastructure components;
- SWIM infrastructure and profiles;
- aeronautical information exchange;
- meteorological information exchange;
- cooperative network information exchange;
- flight information exchange.

a few sectors—not to mention entire ecosystems. ED-201A, attempting to establish an aviation cybersecurity framework across stakeholders, may form part of a possible solution becoming available later this decade. However, in its current approach—until at least a European legislation is in place to support it (expected not before the end of 2021), not to mention US legislation—it would just serve to highlight this concern, as it calls for practical, voluntary sharing and cooperation for the time being.

Even when an updated ED-201A and an imminent RTCA DO-equivalent are in place in the early 2020s, they would still require mandatory, not just voluntary, definitions of domain boundaries, cross-boundary responsibilities, and handshaking. They would also require some form of regulatory responsibilities definitions to go with them, which may become a greater challenge than previously foreseen, as the scope and diversity of regulatory organizations and requirements may turn out to be exceptionally wide. This cross-regulatory scoping process may prove to be a daunting task into the 2020s.

## Lacking Mandated Cooperation Among Stakeholders

As previously discussed, handling cross-domains or even cross-organizational cyber-threats might prove challenging; however, the current ED-201 and planned ED-201A with its imminent RTCA equivalent attempt to resolve at least part of that issue by mandating "External Agreements" with potential partners. These agreements should facilitate proper cooperation in information-sharing, risk-sharing, conditional disclosure, and so on.

However, this approach still leaves a great portion of risks and threats outside the regulatory scope, as it does not go on to mandate "minimum performance" for the proper conduct of aviation stakeholders, such that even stakeholders that do not share agreements would need to disclose minimal sets of information regarding cyber-attacks/threats/risks, for instance. In cybersecurity terms, the current approach actually serves as part of a road map for cyber-criminals for bypassing organizations' cybersecurity in the most efficient way: through attack paths that do *not* run through organizations engaged in agreements with the attacked organization, as chances of such a nonagreement organization voluntarily informing an attacked organization beforehand of a vulnerability related to it would be slim to none.

Alleviating this concern is simple in principle, yet it goes through a political minefield: make the next versions of ED-201 and its RTCA counterpart "minimum performance" oriented rather than "agreement oriented" documents and make the pair "ED-201x" and "DO-xyz" mandatory, rather than voluntary, through proper regulation and legislation. Alternatively, such "minimum performance" aspects could be detailed within related appendices, technical standards by G-32, and/or ACs (FAA) and AMCs (EASA).

## Insufficient Regulatory Progress Outside Commercial-Passenger-Aviation Sector

When discussing aerospace, we need to recall that almost all of the current and emerging cybersecurity regulations and standards are directed at commercial air transport—mainly passenger aircraft, large or small, fixed-wing or rotating-wing—but cargo aircraft regulation is almost always a subset of passenger aircraft regulation. As the military aviation sector increasingly draws from the civilian aviation sector's regulation and standards, we are already witnessing defense/aerospace contracts that mandate the DO-326/ED-202 set. However, other aerospace subsectors are still far behind, as far as cybersecurity regulation and standards; the most prominent now seem to be the space sector, the unmanned aircraft systems (UAS) sector, and the "next big thing" (i.e., UAS traffic management (UTM)). Out of the three, the UAS sector is making modest progress at EUROCAE and RTCA, mainly within a subgroup of the remotely piloted aircraft system (RPAS) committees: WG-105 and SC-228, respectively. The other two are still far behind and would benefit from getting on board with current regulatory/standard-making efforts, or initiate their own efforts.

## Lacking Regulatory Harmonization with Non-Aviation Ecosystems

Outside of civilian aviation, there are many other ecosystems with similar characteristics that are troubled by cyber-threats and are working toward proper regulation: automotive, marine, healthcare, and so on. A partial sample of such sectors that are interconnected with aviation can be seen in Figure 10. As of 2020, there are no notable initiatives to combine cybersecurity regulation for multiple ecosystems, except for some secluded topics. One of the few such initiatives that are already active is SAE International's G-32 "Cyber Physical Systems Security Committee," which is developing JA7496 for Security Engineering Plan, JA6678 for Software Security Assurance, and JA6801 for Hardware Security Assurance, and covers a few ecosystems, mainly aerospace and automotive.

It is advisable for all aviation cybersecurity rule makers and standard makers to adhere as much as practical to the most broadly accepted best practices and frameworks, such as NIST SP 800-82 [35] for Industrial Control Systems, NIST Cybersecurity Framework [36], and other such "gold standards," in order to best prepare for cross-sector cooperation in the future (and make such efforts public). Examples for such efforts of mapping regulatory standards and terms from different sectors are by ENISA [37] in the EU and by the World Bank Group [38], Financial Stability Institute [39], BCG [40], and others in the financial sector.

**FIGURE 10.**   **A few of the sectors interconnected with aviation.**

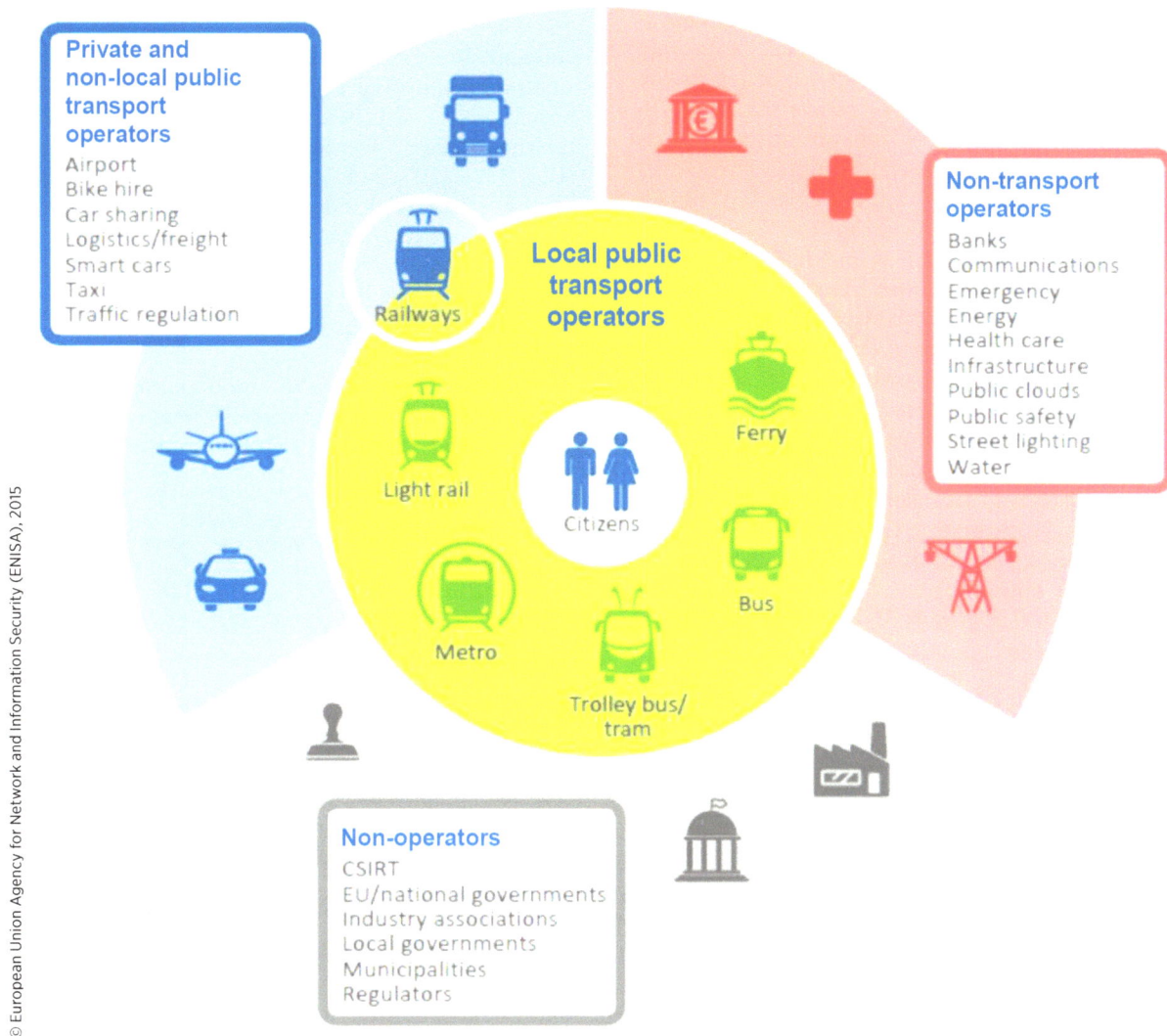

However, as promising as such initiatives may be, across-the-board joint regulation is far from settled. So, any cybersecurity solutions' suppliers would still need to face multiple regulatory systems, raising concerns about the business viability of keeping different hardware/software solutions for different ecosystems. Aviation may be the sector that suffers the most, especially with airworthiness aspects, as the absolute numbers of items are dwarfed by other sectors, such as automotive.

## International Mistrust/ Distrust: "Which Side Are You On?"

The organization vested with international authority for regulating the entire global aviation sector is ICAO—as nearly all UN members are ICAO members. As such, ICAO's importance in regulating aviation cybersecurity is paramount, and resolution A40-10, taken at its 40th Assembly [41], held in 2019, addresses "Cyber-security in Civil Aviation," and mandates essential steps, carefully designed beforehand by SSGC [42] (the product of previous resolution A39-19 of the 39th Assembly [43]).

However, as professionally sound and helpful the content of resolution A40-10 is, one concern looms large: among the prominent state members of ICAO, there are a few that were quite recently related to cyber-attacks on civil aviation domains [44, 45]. In fact, the original language proposed for this resolution called for an "aviation *trust* framework" [46]; however, the resolution itself only states "cyber-security framework." "Trust," one of the most essential attributes of cybersecurity, is apparently nowhere to be found.

Whereas safety does not rely upon trust—just science, engineering, and complying with regulation—for security, especially cybersecurity, "Trust" is paramount. It is the "underpinning" of the DO-326/ED-202 set documents, as well as any other aviation cybersecurity regulation. But if untrustworthy actors might, by default, be part of the only viable international regulatory organization (i.e., deep inside aviation's "security perimeter"), some concerns might be in place, which would not be very different from concerns arising a few decades ago for kinetic-terror (e.g., Would all members actually comply with such regulation? Would all members even want to comply? Could any members actually abuse such regulation? Could any members even influence the regulation-making process other than in bona fide?)

Furthermore, some prominent organizations [47] rightly recommend to "define and incorporate a more universally applicable use of trusted and untrusted actors within the aviation ecosystem," but what if some of the actors actually developing the rules are untrustworthy? In fact, these untrustworthy actors, that are defined by the US National Security Agency's (NSA) Information Assurance Technical Framework [48] to be the highest Threat Levels—namely T6 and T7: "Extremely sophisticated adversary with abundant resources […] e.g. nation states […],"—are the mythical "elephant in the room" (Figure 11).

## Recommendations

The aviation ecosystem is marked by a fractured, multitiered, and multispeed types of stakeholders, from nation-states to small private businesses, in the aviation sector and other related sectors. This domain is mainly in the hands of large international or government organizations such as ICAO (UN), FAA (US), and EASA (EU). However, there is some room for technical industry organizations, such as SAE International (through G-32), RTCA (through SC-216), EUROCAE (WG-72), and others, to facilitate these large global undertakings by feeding such large organizations with useful inputs and standards:

- Filling more standardization gaps for aviation subsectors currently (almost) unattended, such as ATM/ANS (outside Europe) or airports (everywhere)

- Setting clear domain borders and responsibilities among the various stakeholders, along the lines of the DAH/operator distinction of DO-355/ED-204

- Setting "minimum conditions" for cooperation across stakeholders' and sectors' responsibilities,

**FIGURE 11.** "The Elephant in the Room."

"I suppose I'll be the one
to mention the elephant in the room."

**FIGURE 12.**   **Visualization of commercial aviation with supply chains.**

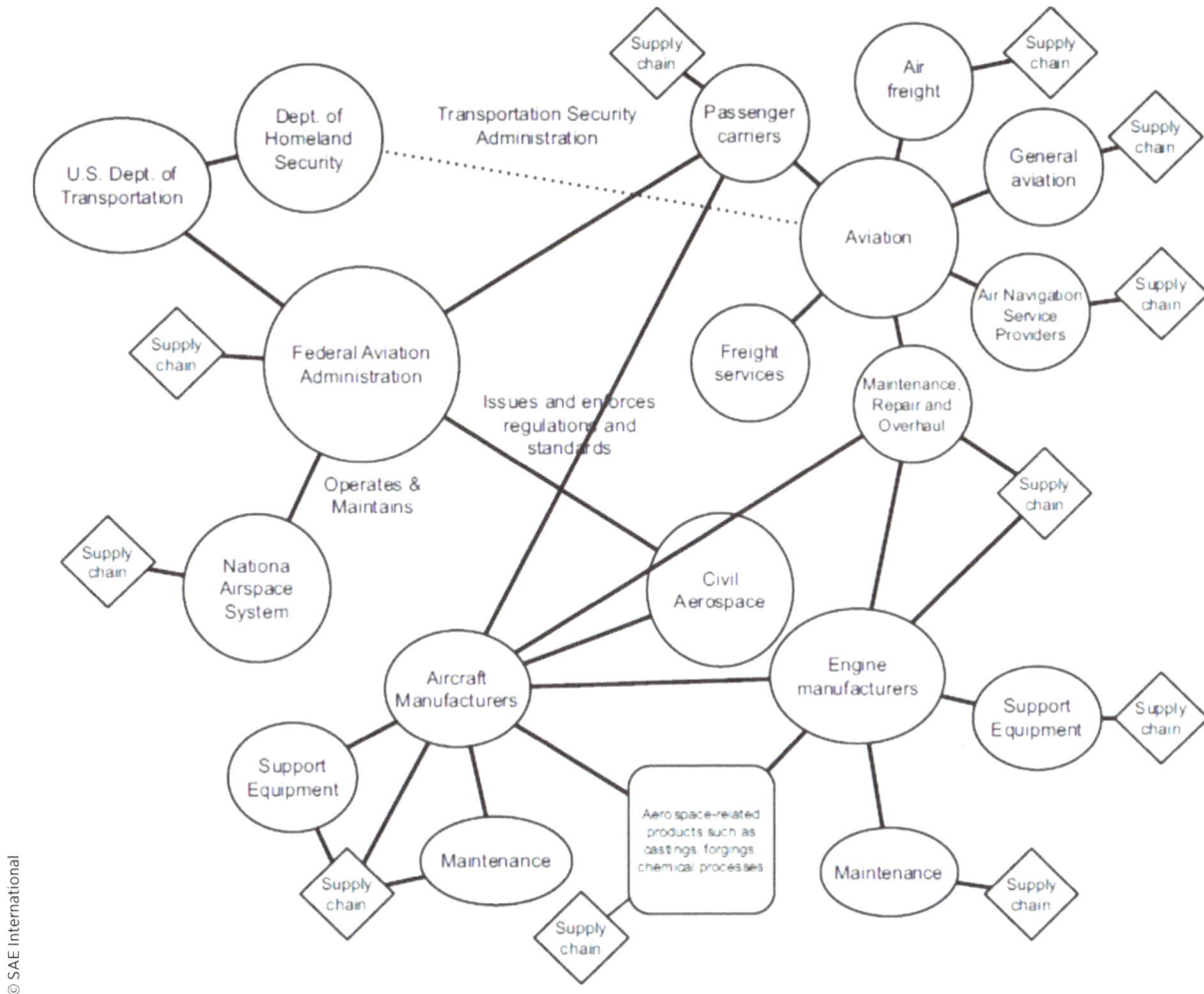

© SAE International

including properly mandated mutual disclosure and assistance-setting a mandated lower bar for external agreements

- Enhance cross-sector cooperation such as G-32 shared aerospace/automotive endeavors

- Strengthen the cooperation of cybersecurity standardization efforts and specific subsectors standardization efforts (e.g., deeper cooperation of SC-216/WG-72 with SC-228/WG-105 for UAS)

- Develop concrete criteria for establishing "Trust," even for the process of regulation-making, in order to enable international organizations to make practical steps toward a "trust framework"—a potentially viable option to promote such a trust framework could be voluntary initiatives by nongovernmental entities like the World Economic Forum (WEF) "Cyber Resilience Playbook for Public-Private Collaboration" [49] or commercial initiatives like the "Charter of Trust" [50]

# OEM Supply Chain Issues

Aircraft development and production is an extremely complex process, involving a large number of suppliers, starting with the original equipment manufacturer (OEM) at the top, proceeding through the supply chain through Tier-1 suppliers, down to the suppliers of the simplest items, literally "the nuts and bolts." As of 2019, Airbus reports having more than 12,000 suppliers worldwide [51] and Boeing more than 20,000 [52]. Such gigantic aviation supply chains (Figure 12) have become prime targets for cyber-attackers [53].

Regulating the safety and security of such enormous aviation supply chains brings up cybersecurity issues that are yet to be settled, some of which are already recognized by key industry actors [54], while others are still elusive:

- Suppliers' uncertain compliance capability

- Hardware cyber-threats

- Legacy/COTS items
- Insufficient disclosure upstream
- Scoping in/out—"Mass Extinction?"
- The internal supply chain

# Suppliers' Uncertain Compliance Capability

Airworthiness cybersecurity regulation and standards encompass, as can be expected, the OEMs as well as their entire supply chain. There are even specific separate provisions in the DO-326/ED-202 set text separately related to the distinct hierarchical levels of aircraft, systems, and items-mainly in DO-356A/ED-203A so that the entire supply chain should be adequately covered.

However, although these efforts seem, at first glance, quite adequate for the various supply chain actors, in reality, the task might turn out to be daunting for the lower tiers of the supply chain, especially for the smallest subcontractors (in particular, those which are not yet accustomed to developing and/or producing safety-critical items).

One obvious reason for this adaptation challenge is the stricter practices mandated by the new cybersecurity requirements, which may, in certain cases, overwhelm the smaller actors. Another reason for this issue may be the fact that the OEMs and the higher-tier organizations were the most prominent industry representatives in many of the committees and subcommittees that developed the DO-326/ED-202 set, thus, naturally, not being fully attentive to the smaller actors' woes.

A possible mitigation for this mismatch might be for the OEMs and higher-tier actors to step in and become more involved with their supply chain than they currently are, which might destabilize the delicate existing balance of some supply chains. However, for this to materialize, even beyond the scope of ED-201 (or even the future ED-201A and its RTCA counterpart), an extension of the ED-201 risk management framework is advisable—potentially along the lines now developed by SAE's G32, combined with such resources such as ISO27036 [55], NISTIR 7622 [56] plus the NIST Cyber Supply Chain Risk Management [57] in general, the US FY 2020 NDAA [58], the DFARS "Disclosure of Information" guidance [59], and others.

# Hardware Cyber-Threats

When approaching supply chains from a cybersecurity aspect—for aviation or any other sector—there would typically be two types of tangible deliverables to consider, including their support, documentation, and so on: software in any shape or form and hardware in any shape or form, and this report would not even attempt to properly mark the borderline between the two. However, when the DO-326/ED-202 set is carefully examined, it is entirely obvious that software is very thoroughly treated (although specific unsettled topics concerning such

software will later be discussed in this report), whereas hardware is almost an afterthought. Hardware is definitely not thoroughly treated, except for Complex Electronic Hardware (CEH), which can almost be regarded as a different form of software, with only some subtle variations to set it apart from pure software. Even when hardware is shortly discussed, it is only electronic hardware—not mechanical entities, such as the shape of engine or rotor blades that can be hacked [60]. One major factor in the difficulty to apply cybersecurity against hardware-related attacks is the prohibitive nature of quickly patching a solution, if and when an attack of this nature is revealed—a most common solution for software security updates.

A possible reason for this disparity could be the scant background and available material concerning hardware aspects of cybersecurity. Unlike software aspects, hardware cybersecurity is an evolving discipline and not very mature. The one hardware aspect that is lightly touched upon is securing electronic hardware against counterfeit [61], but again there is just a brief discussion with very few top-level recommendations. The implication of such vague standards and regulations is a clear and present cybersecurity issue for supply chains.

Pure hardware aspects would definitely add to any appropriate regulatory discussion such topics as logistics, physical delivery, and so on—with all their cybersecurity-related issues (e.g., non-bypassing/non-tampering assurance). But even combined software-CEH hardware, cybersecurity aspects may already be developing due to the emergence of the system-on-chip (SoC) concept, which blends such elements as central processing units (CPUs), memory, graphics, physical connectivity, and in many cases, Wi-Fi capability. SoCs pose an overwhelming combination of security risks, including unspecified hardware and hard-to-detect backdoors, and may incorporate elements from various vendors, less documented compared to suppliers of cards, and other risks. Such SoC security issues are already presenting themselves [62].

A possible path to start solving this issue is, again, SAE's G-32, which already has hardware aspects of cybersecurity on its agenda with the development of JA6801. However, it may potentially require the revisiting of at least DO-356A/ED-203A for hardware aspects, preferably adopting the upcoming products of G-32's HwA (Hardware Assurance) subgroup, and basing decisions on solid resources such as SAE 5553 [63], emerging initiatives such as Trusted and Assured Micro-Electronics (TAME) [64], and applicable studies [65, 66]. This standard-making process can also draw from such existing nonaviation-specific supply chain practices and handbooks, such as the "Software Safety and Security" section of International Aerospace Quality Group's Supply Chain Management Handbook (SCMH) or similar. The end result of G-32 can be expected to be a combined system-hardware-software security framework, along the lines of Figure 13.

# Legacy/COTS Items

Even assuming that the hardware regulatory infrastructure and standards were in place for cybersecurity, legacy and/or commercial off-the-shelf (COTS) hardware issues still loom

**FIGURE 13.** Cyber physical systems security framework.

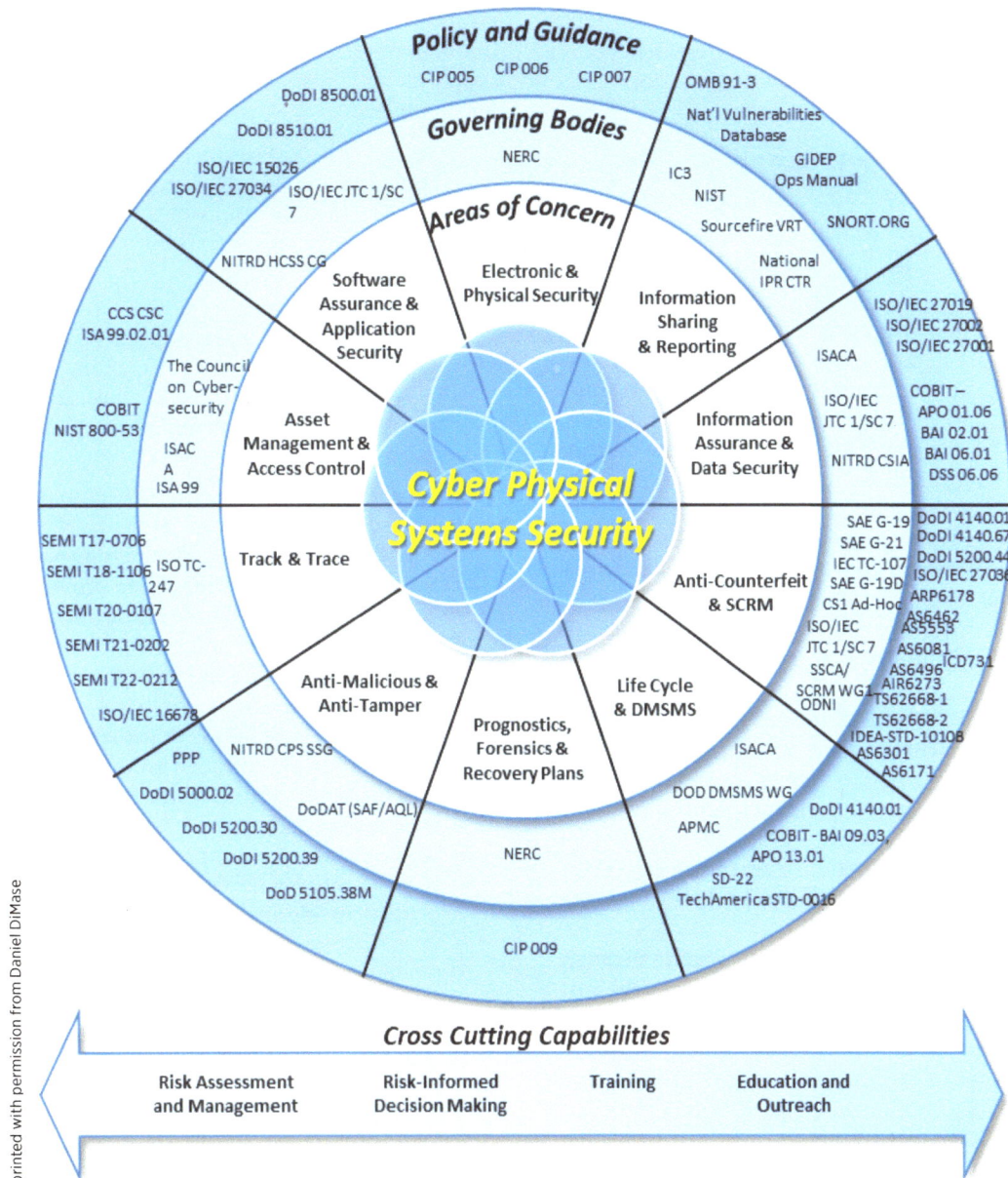

large. Although DO-356A/ED-203A do provide some certification advice for such hardware (and even software), it only applies to new or modified aircraft planning to use such hardware.

The deeper issue, though, lies elsewhere: as current legacy airborne systems and/or items become obsolete, every modification, even one that would have been deemed "minor" before the era of cybersecurity regulation, stands relatively high chances of being deemed "Major" due to only cybersecurity, if nothing else. How so? For instance, electronic chips that were certified in the past (earlier than 2020) for aerospace use, were considered high-quality, safety-critical components, may now be deemed vulnerable to cyber-attacks, or

maybe even being exposed as actual cyber-threats at hindsight [67], so any risk assessments performed under the DO-326/ED-202 set would rule out their usage on-board any commercial airplane.

For legacy aircraft or systems, such an occurrence can wreak havoc along supply chains, forcing unplanned "urgent" changes—only due to the passage of time—eroding the security level of existing systems and items.

There are no easy solutions for such scenarios, as gradually, entire aircraft supply chain branches may turn in one of two directions: either keep pre-cybersecurity procedures, thus excluding this aircraft type, or at least the specific branch in question, from aviation's "security perimeter" ("scope out"); or entering the uncharted territory of retroactive regulation, at least partially. Mandatory periodic risk assessments, discussed later in this report—with gradual treatment of the findings—may somewhat alleviate such implications.

# Insufficient Disclosure Upstream

With such odds, one of the possible mitigation strategies for supply chain issues, as previously mentioned, is for the OEMs to get deeper than they currently would (or would even prefer) into suppliers' details, and assist lower tiers with their new cybersecurity certification issues—or else "weed out" those who might pose security weaknesses, vulnerabilities, or even threats for the entire aircraft. In fact, the current DO-326/ED-202 set mandates the submittal of a System Security Integration Guidance (SSIG) from item makers to system makers and from system makers to OEMs, which provides for downstream disclosure without any mandatory upstream parallel.

However, in order for such upstream assistance to be effective, OEMs would need to actually disclose, on a need-to-know-basis, a fraction of their own proprietary information as well, so as to allow for adequate security environment assessment for their trusted suppliers. Not only is active assistance up the supply chain uncommon in today's business conduct, but the actual disclosure of proprietary information upstream—although practiced on a small scale through nondisclosure agreements (NDAs) for interfaces and such—is almost completely the opposite of current practices.

The most probable outcomes for insufficient upstream disclosure are either more iterations (with time and cost adding friction to existing aviation supply chain) or a complete paradigm shifts. An obvious mitigation would be the introduction of an "upstream SSIG" in the scope of ED-201 and/or DO-356A/ED-203A, with at least silent consent from the large OEMs. Such consent can enable the OEMs to influence the outcome in the direction of minimal exposure with maximal benefit to their supply chain.

# Scoping In/Out—"Mass Extinction?"

As was previously discussed, aviation supply chains may soon face various challenging issues as OEMs are already trickling down the new cybersecurity regulatory guidelines in the form of requirements and specifications. The previously described resulting scenarios have one quality in common: paradigms are going to shift, this way or another, as such new requirements and specifications will naturally push many suppliers beyond their current comfort zone.

As always, when paradigms shift, there are going to be quite a few actors that cannot—or will not—shift with the rest of the industry. The phenomenon that may take place in the case of aviation cybersecurity certification might be the creation of a new chasm between those that adjust to the imminent paradigm shift and those that do not. This new supply chain landscape would present three major options for suppliers:

1. Secure their systems/items against relevant cyber-threats to a reasonable level, leading OEMs to regard them as "Secure" and—in some cases—even define their systems/items as "Security Measures;" in other words: *"scoped in."*

2. Remain with unsecure systems/items that only meet their functional requirements, leading OEMs to regard them as "Unsecure" or even potential "Security Threats," thus excluding those systems/items from their aircraft; in other words: *"scoped out."*

3. A practically balanced option for mature systems/items that are too essential to do away with ("assets"), yet too delicate to go through more than light updates would be to provide evidence, or even perform light modifications, so that the system/item would at least not be assessed as a cyber-threat, even if it is not properly protected (Security Assurance Level-0 (SAL-0) or even SAL-1), so as to allow OEMs to incorporate them under the auspices of their own security architecture and measures. In other words, the critical system/item would be *"scoped in" for a reasonable cost* to the OEM for extra protection. This approach would probably be a reasonable choice for most critical, rarely available subsystems (e.g., sensors or actuators), but probably not for abundant, somewhat less-critical types of systems—especially systems that reside in the "periphery" of the avionics (e.g., most communication systems) that will still need to make tougher choices.

Although this may seem like mainly a suppliers' issue, OEMs may also find themselves between a rock and a hard place, as—in certain cases—all of their options for certain crucial systems/items would be "scoped out," so that they may face some hard choices in order to comply with regulation. Those choices would be to assist the providers of such systems/items, develop/produce them in-house, or even change higher-level system design to do away with the requirements for these systems/items.

OEMs and suppliers would be well advised to immediately start developing their strategic decisions and even start implementing such decisions, rather than wait for this regulatory earthquake to split the ground beneath their feet. One way to approach this issue is to designate neutral, certified mentors, trainers, and so forth to support suppliers, so "Trust" would only need to be established once, as OEMs would be at least partly relived from the need to actively support suppliers while suppliers would have access to immediate, practical "life-saving" support.

# The Internal Supply Chain

The last, but by no means least, aspect of aviation supply chains to be discussed might seem awkward at first glance: the top of the chain—the OEMs themselves. Surely, as OEMs are very much aware of aviation cybersecurity challenges, and took a very (very) active part in crafting the new regulation—including the DO-326/ED-202 set itself. It could have been expected that the OEMs would be the gold standard of aviation cybersecurity regulation.

But truly, although OEMs deploy some of the latest-and-greatest cybersecurity systems in the aviation sector, surprisingly (or not?), the new regulatory framework has next to nothing to show for internal development/production processes pertaining to cybersecurity. Unlike the products at stake—aircraft/systems/items, software/hardware—for which, notwithstanding unsettled topics, regulation is either in place or about to be in place, there is almost no regulatory discussion whatsoever of cybersecurity aspects for internal processes of the organizations developing and producing these products (OEMs, but also system makers of the higher tiers). The relatively mild internal/external approach of ED-201 is still a long way from becoming a mandatory document, and the very short "tool security" subsection of DO-356A/ED-203A is still too slim and too vague to make a difference. So, for the time being—even OEMs are facing long odds in coping with cyber-threats for their internal development and production systems [68].

The obvious way forward for approaching this issue is to have it addressed by the next editions of the DO-326/ED-202 set and/or within the scope of G-32, as is also recommended by the Aerospace Industries Association (AIA) [69].

# Recommendations

Large aviation supply chains are arguably the richest source of cyber-threats for OEMs and the entire aviation ecosystem. Overwhelmed suppliers, uncooperative OEMs, inherent hardware trust issues (especially with COTS and legacy items), and even the OEMs development/production process ("internal supply chain") issues—coupled with the entire domain being almost an afterthought in current regulation and standards—pose real cyber-challenges. Consequences might be serious, from an increased burden on OEMs to dramatic "Mass Extinction" of complete supply chain branches and not all of the associated issues can necessarily be mitigated by proper regulation (or at all). Some potential solutions may be

- Follow-up the DO-356A/ED-203A security assurance of systems and items with specific transition criteria from OEMs to suppliers (e.g., requirements) and vice versa (e.g., products). The current DO-201A development efforts include some beginnings—that are advised to be enhanced and later be made mandatory by regulation.

- Considerably enhance the hardware-specific content of the DO-326/ED-202 set, by adopting such upcoming inputs from G-32.

- Carefully consider the implications of existing COTS and legacy equipment already embedded in the security perimeters of aircraft and devise an informed policy to handle them—this should be a joint effort by regulators, industry, and standard makers, and it is not yet clear how such an outcome might even look like.

- Correct the asymmetrical disclosure mandates, whereas suppliers are obliged to provide OEMs with a SSIG, OEMs are obliged to … well … nothing in particular.

- Start a "Mass Education" of suppliers on the upcoming regulation, even by OEMs if necessary, to attempt to mitigate the risk of "Mass Extinction."

- Considerably enhance the "tools security assurance:" sections of DO-356A/ED-203A and of the DO-326/ED-202 set, to address the cybersecurity of the development process itself—even consider allocating a separate DO/ED, similar in nature to the DO-178/ED-12 set's DO-330/ED-215, "Software Tool Qualification Considerations" [70].

# Regulatory Compliance Challenges

As could be expected, any new regulation, especially when airworthiness is at stake, takes time to settle and reach stable equilibrium. Airworthiness cybersecurity might be especially challenging, as the erratic and fast-paced inherent nature of cyber-threats meets what is arguably the most conservative regulatory discipline for many decades: airworthiness. Major expected challenges, for applicants and regulators, may include the following:

- "Regulation is not profitable!" + "Cybersecurity is not profitable!" = ?

- Conflicts with other regulatory frameworks

- The Safety and Security "Rookie Wall"

- Latent safety prerequisites

- How deep should regulatory documentation go?

- How deep should regulators go?

## "Regulation Is Not Profitable!" + "Cybersecurity Is Not Profitable!" = ?

Airworthiness regulation and cybersecurity are always expensive; they both involve a priori investments in order to prevent a posteriori incidents. The combination of both should be extremely costly, thus not highly motivating applicants to comply unless they absolutely have to. The outcome might be applicants' compliance-oriented approach (i.e., shooting for the minimum level of compliance, regardless of the actual airworthiness cybersecurity considerations).

There is no simple solution here—probably not even the "Black Swan" [71] theory conclusions that should persuade stakeholders to invest a reasonable amount in order to avoid incidents that may bring down their business altogether: similar to insurance, but preventive rather than responsive. Mandating insurance, such as for automobiles, may serve as a mitigating factor by itself—as emerging cyber-insurance policies [72] routinely include the level of regulatory compliance for a premium discount, so as to lower insurers' risks— thus providing a financial incentive for the insured party to certify. Such an approach is already successfully applied in other sectors (e.g., the energy sector for micro-grids [73, 74]).

Another approach applied in the energy sector, by the North American Energy Reliability Corporation (NERC), is the enforcement [75] of NERC Critical Infrastructure Protection (CIP) standards [76], resulting in heavy fines for violations [77]: a "stick" to match insurance-discount's "carrot." Although this approach may break the backs of small businesses, it may also move the profitability equation toward more adequate cybersecurity regulatory compliance.

It may also be helpful to educate as many stakeholders as practical regarding the cost, in hard cash, of such incidents, versus the cost of applying proper cybersecurity practices, along the lines of the aviation cybersecurity regulation/standards/best practices, so as to push the perceived net cost even lower [78].

## Conflicts with Other Regulatory Frameworks

Before even embarking on the process of airworthiness certification for cybersecurity, there might be an inherent conflict out-of-the-gate with other sets of legislation and regulation, due to the conflicting nature of such sets. This, by no means intends to point out any bureaucratic or tactical conflicts, but rather an inherent conflict, embedded within the set of conceptual priorities of various regulatory sets, serving different ecosystems, as stressed before. Most other existing sets of cybersecurity regulation and even legislation (e.g., General Data Protection Regulation (GDPR) [79]) focus on privacy or—at a minimum—emphasize the "confidentiality" attribute of cybersecurity. On the other hand, airworthiness cybersecurity regulation focuses on safety, thus emphasizing the "availability" attribute of cybersecurity.

This mismatch is so fundamental that it is destined to produce imminent conflicts with other regulation sets if not properly handled. In fact, DO-356A/ED-203A explicitly puts "fail-safe" before "fail-secure," which is exactly the opposite of some existing cybersecurity regulation.

The most reasonable mitigation for such a regulatory conflict would be for airworthiness regulation-producing organizations to tighten cooperation with other sectors' (potentially conflicting) cybersecurity regulation-producing organizations. A more stable solution for the long run should be to define clear priorities in national legislation and an international umbrella regulation, which may last at least a decade or two.

## The Safety and Security "Rookie Wall"

One major issue that can be already expected in the aviation cybersecurity regulatory process is what can be called "The Rookie Wall:" the first encounter of applicants lacking previous background in safety-critical systems certification with a process of yet-unknown qualities for them. This phenomenon can be expected to mostly hit the makers of systems such as IFECs and EFBs, which (until now) were not conceived to be highly safety critical by themselves, but can now be potentially exploited to become attack vectors or attack paths' elements. Furthermore, a common misconception regarding such external systems calls "to incorporate strong security mechanisms in lower cost system [80]," applying low SALs as a low-cost method to substantially harden large systems, disregarding the fact that low SALs are for relatively light security measures, while higher SALs require a lot of the higher Design Assurance Level (DAL) safety-assurance measures—worsening the rookie effect. Whereas rookies, mostly IT systems developers, might be security-savvy, hence quite open for the DO-326/ED-202 set basic concepts, safety might be an entirely new discipline for them, as they would need to cross the fine line into cyber-physical systems domain, where safety-critical systems are a subset of security-critical systems (Figure 14). However, safety engineering and security engineering only partially overlap (Figure 15), and security development is integrated with safety development.

Such system makers should be made aware of this issue as early as practical and would be best advised to not only educate themselves with the new DO-326/ED-202 set, but start with the set's foundations: the basic concepts of safety-critical systems certification, as represented by the ARP4754/4761 set [81, 82] and the DO-178/DO-254 set [83, 84] (Figure 16), or else many DO-326/ED-202 set hidden assumptions (Figure 17) might not be properly realized. For the rookie level, it would probably suffice to opt for standard, publicly or privately available DO-178/DO-254 training [85, 86, 87, 88], ARP4754/4761 [89, 90, 91, 92] training, and some already available DO-326/ED-202 set training [93, 94]—separately or as a bundle—which can serve as a shortcut to concepts that took more mature companies decades to integrate in their internal processes. As imperfect as this approach might be, it may somewhat alleviate preliminary pains.

**FIGURE 14.** Relationship between safety-critical and security-critical systems.

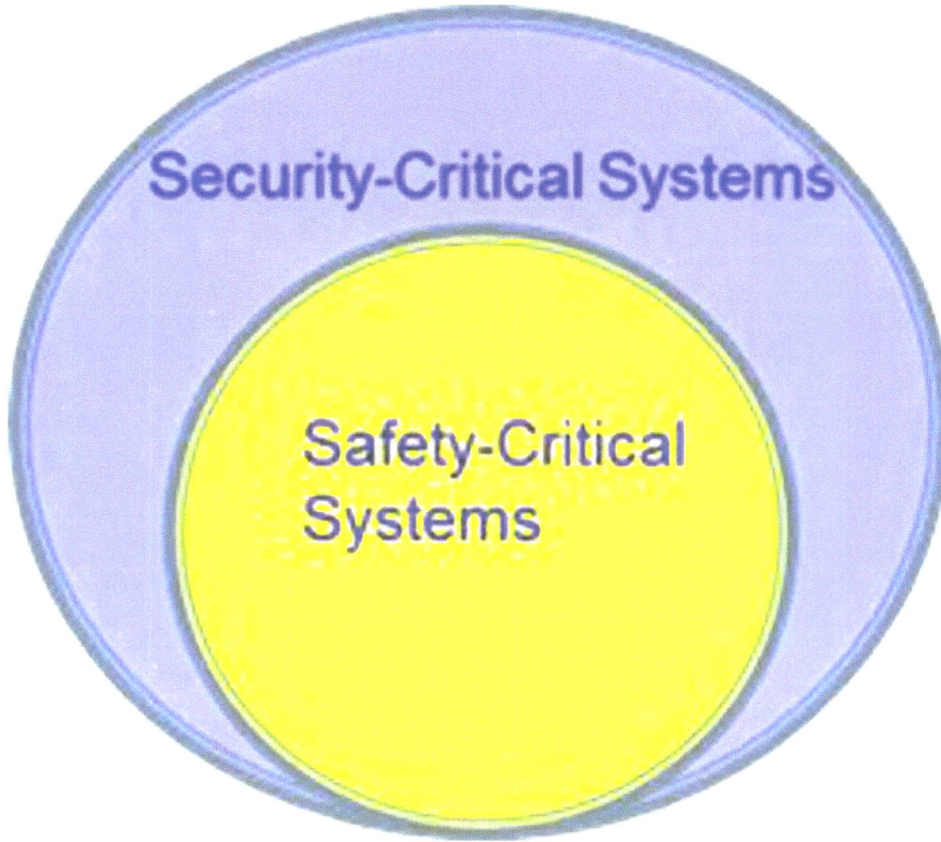

© SAE International.

**FIGURE 15.** Relationship between system safety and system security engineering elements.

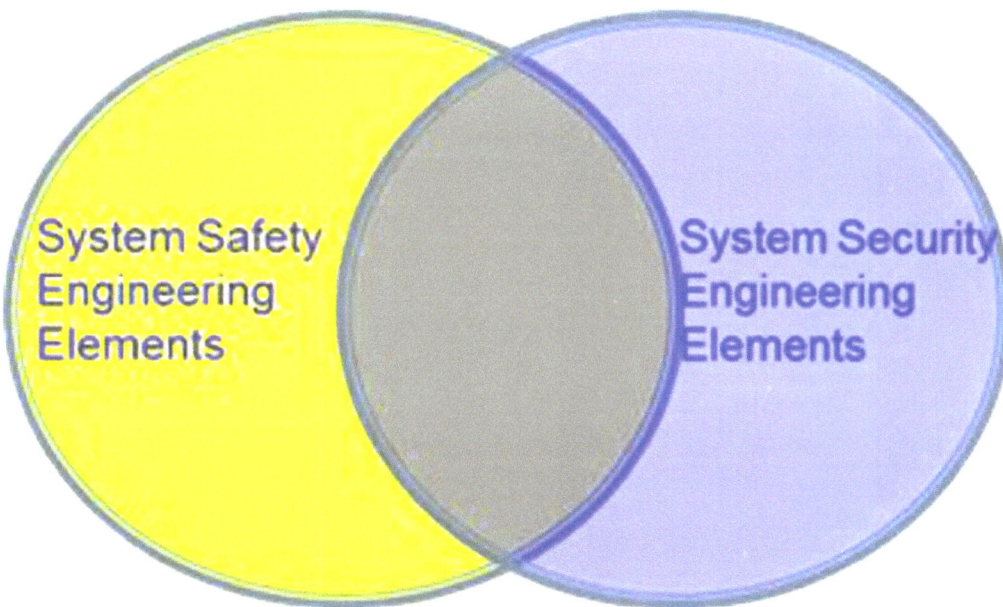

© SAE International.

**FIGURE 16.**   "The Aviation Certification Eco-System™"— pre-2010, "classic" safety.

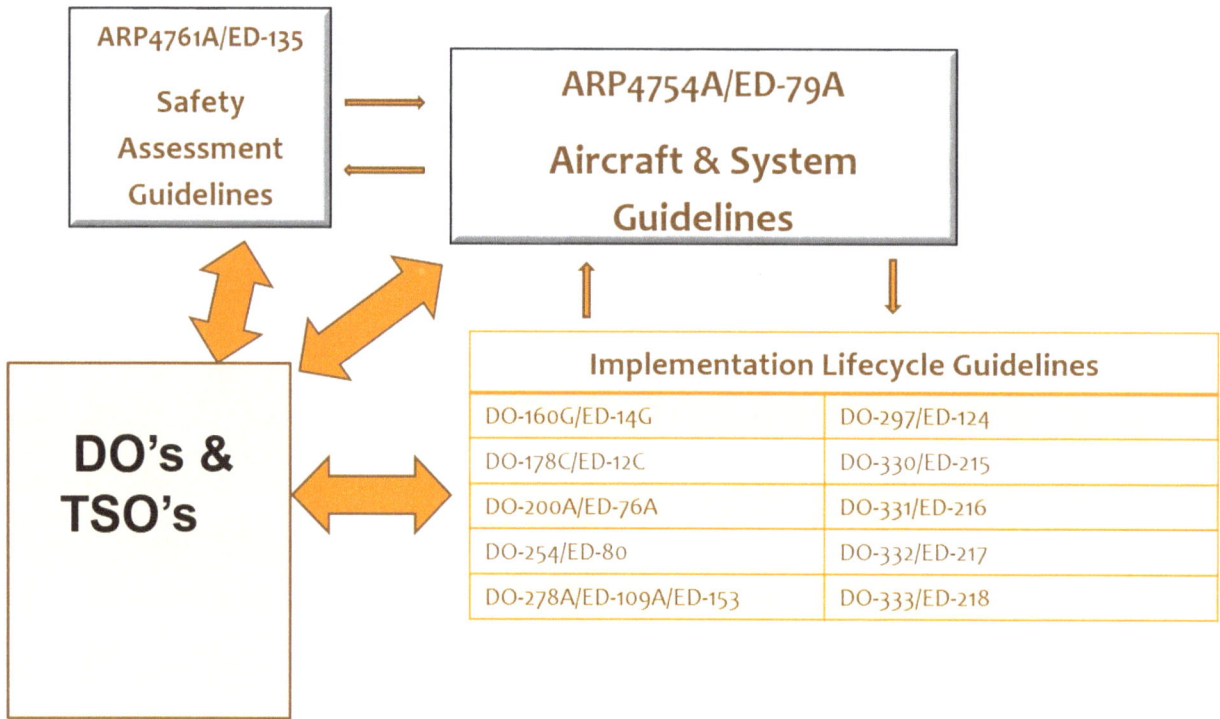

**ARP4761A/ED-135**
**Safety Assessment Guidelines**

**ARP4754A/ED-79A**
**Aircraft & System Guidelines**

**DO's & TSO's**

| Implementation Lifecycle Guidelines | |
|---|---|
| DO-160G/ED-14G | DO-297/ED-124 |
| DO-178C/ED-12C | DO-330/ED-215 |
| DO-200A/ED-76A | DO-331/ED-216 |
| DO-254/ED-80 | DO-332/ED-217 |
| DO-278A/ED-109A/ED-153 | DO-333/ED-218 |

© AFUZION-InfoSec

**FIGURE 17.**   "The Aviation Certification Eco-System™"—post-2018, including the DO326/ED-202 set for security.

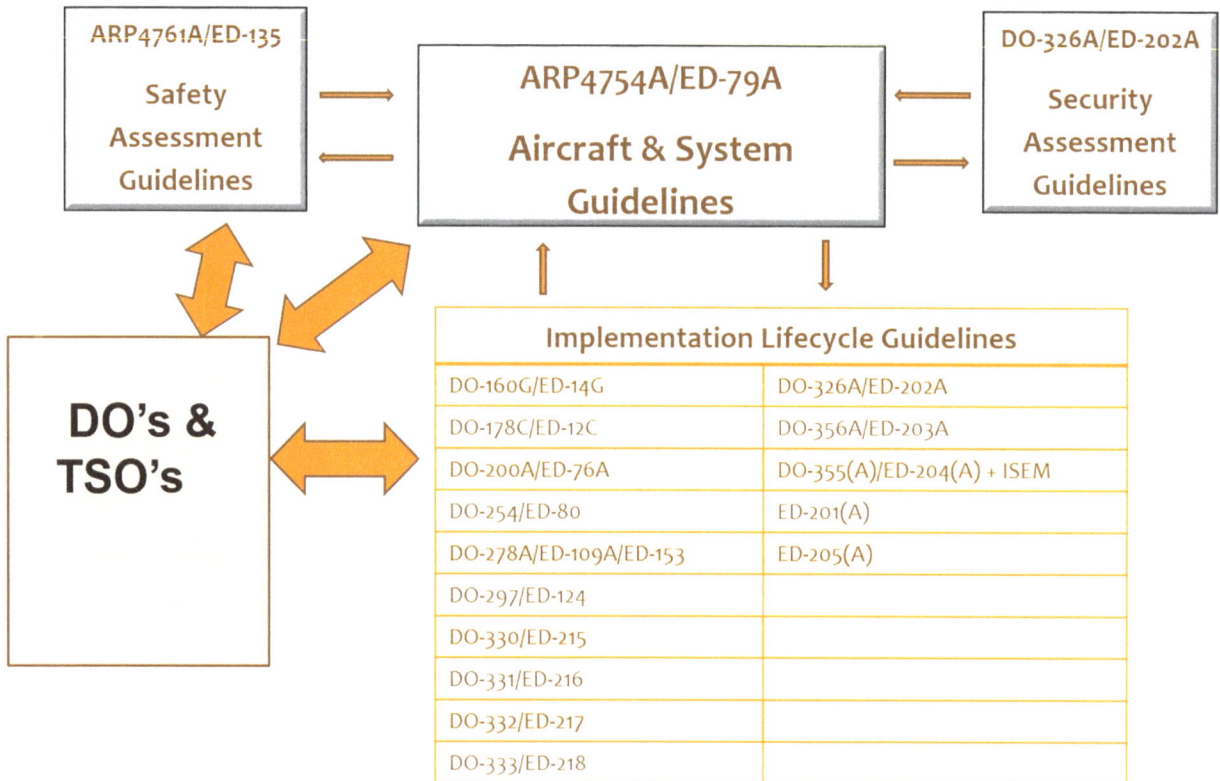

**ARP4761A/ED-135**
**Safety Assessment Guidelines**

**ARP4754A/ED-79A**
**Aircraft & System Guidelines**

**DO-326A/ED-202A**
**Security Assessment Guidelines**

**DO's & TSO's**

| Implementation Lifecycle Guidelines | |
|---|---|
| DO-160G/ED-14G | DO-326A/ED-202A |
| DO-178C/ED-12C | DO-356A/ED-203A |
| DO-200A/ED-76A | DO-355(A)/ED-204(A) + ISEM |
| DO-254/ED-80 | ED-201(A) |
| DO-278A/ED-109A/ED-153 | ED-205(A) |
| DO-297/ED-124 | |
| DO-330/ED-215 | |
| DO-331/ED-216 | |
| DO-332/ED-217 | |
| DO-333/ED-218 | |

© AFUZION-InfoSec

## Latent Safety Prerequisites

The Rookie Wall should have been relatively easy to predict—in fact it was predicted long ago, including some subtle implications in the DO-326/ED-202 set text. But even more mature organizations may yet encounter some unforeseen challenges with this new process, even at their core competencies of safety-critical systems certification. One such issue would most likely relate to one of the most crucial foundations of the DO-326/ED-202 set, namely the ARP4754/4761 set of system engineering and safety (Figure 18)—including some major parts of the DO-178/ DO-254 set.

Although the ARP4754/4761 set has long been the de facto system development and safety acceptable means of compliance for airworthiness, it was not yet made official through (for instance) 14CFR part 25, hence its implementation is not at a uniform level of compliance throughout the aviation industry, to say the least. Furthermore, the forthcoming revision-B to ARP4754A and revision-A to ARP4761 mandate increased aircraft-level safety coordination, which is inclusive of aircraft-level cybersecurity. As a result, although many of the mature actors in the industry do follow at least the principles of the ARP4754/4761 set, in many cases, actors beyond OEMs and Tier-1 are even less inclined to follow such principles. This results in some meaningful gaps, mainly in the documentation baseline.

These gaps are about to become a real challenge for even relatively mature organizations, as the ARP4754/4761 set requirements are present all over the place in the DO-326/ ED-202 set, directly and indirectly, so that ARP4754/4761 are—for all practical purpose—prerequisites for cybersecurity certification. Such actors should be advised to start closing their ARP4754/4761 gaps immediately, so as not to face uphill struggles even before reaching the cybersecurity certification process per se.

## How Deep Should Regulatory Documentation Go?

Regulations and standards and guidance documents are naturally the documents to refer to when embarking on a regulatory certification process. However, the proper level of detail of such documents is still unsettled. Whereas top-level, low-detail documents can be expected from organizations like ICAO and from legislation texts, it is still unclear what level of detail is proper for practical regulatory organizations (e.g., FAA and EASA) or even what level of detail is proper for any industry standards they choose to adopt.

Judging by the history of software and hardware airborne regulation, such as DO-178/ED-12 and DO-254/ED-80, in the long term, this type of regulation tends to settle around the process level, leaving technical details at the discretion of

**FIGURE 18.** Safety-security integrated development with the new DO-326/ED-202 set.

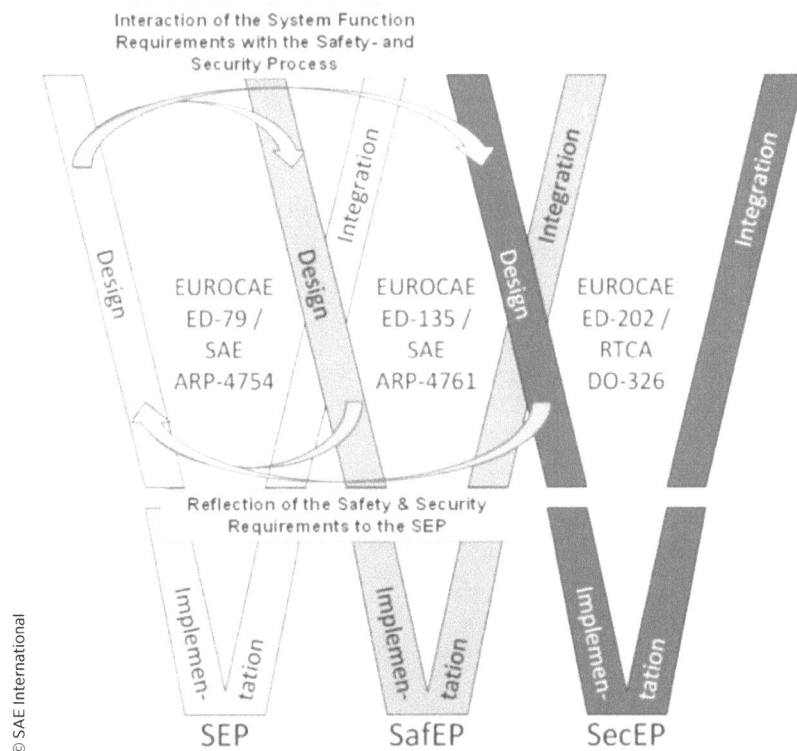

applicants, and the general attitude of the DO-326/ED-202 set may support such a conclusion. However, in the case of cybersecurity (unlike safety) the details, especially those related to system architecture and software, do need to be a bit more specific, as security is not just a general requirement like safety—it is also a specific content domain. On the other hand, going into too much detail, especially in direct regulatory mandates, may pose even greater drawbacks, such as forcing one-size-fits-all-type solutions that would become suboptimal for all, having to chase the rapid pace of cybersecurity developments, and more.

At the end of the day, it would seem reasonable to trim the DO-326/ED-202 set to a process-oriented guidance, while any deeper technical matters would be better served by referring to standards and best practices documents, such as the emerging SAE JA7496, JA6678, and JA6801. Preferably, every higher-level regulation should endorse and refer to the appropriate lower levels, so as to make for a solid regulatory process definition along the lines of the current definitions of "Acceptable Means of Compliance" by FAA and EASA.

## How Deep Should Regulators Go?

An issue of utmost importance for any aspect of airworthiness regulation is the level of expertise inside the regulating organization (e.g., FAA and EASA), which reflects upon the level of involvement of the regulators in the certification process itself. As the airworthiness cybersecurity certification process appears to be evolving, it is destined to involve highly skilled professionals covering a broad spectrum of disciplines: software, hardware, networks, system engineering, industrial processes, cybersecurity, and more. Furthermore, cybersecurity is a relatively young discipline, especially for Cyber Physical Systems (CPS), specifically for the aerospace industry, that until not very long ago considered itself immune from cyber-threats—and it keeps evolving at a very rapid pace to keep up with constantly morphing threats.

Thus, more than the classic, mature, aspects of airworthiness, cybersecurity poses a real challenge to the technical expertise of regulators, as it is not just a relatively new discipline: it also keeps moving faster than classic disciplines. The point of equilibrium for this new regulation is yet to be determined, as it may prove to be virtually impossible to contain the entire spectrum of required expertise within the regulating authorities, partly because of the demand/supply ratio of such experts. However, history proves that extending the authority of Designated Engineering Representatives (DERs)/Design Organizations Approvals (DOAs) too far may weaken required checks and balances.

Eventually, it can be expected that time should tell how such a stable equilibrium would manifest itself, but possible solutions could be building on the cumulative experience of the FAA and EASA with Special Conditions issued for airworthiness cybersecurity over the last 15 years. They could also be relying heavily—more than is common for "pure"

safety—on trustworthy external experts, in order to keep up with a constantly changing cybersecurity landscape, while maintaining strong in-house expertise that would be based on the core competencies of regulating authorities (e.g., certification processes).

# Summary/Recommendations

The certification process itself may (and will) pose its own challenges: the fact that aviation cybersecurity regulation is expected to be even less economical than other types of regulation, conflicting requirements with other existing cybersecurity regulatory processes, the effects of the underlying safety requirements on rookies and established actors alike, and the level of detail and expertise of regulations and regulators. Some of these challenges cannot be easily settled, but better practices and mitigations can be pursued:

- Enhanced education of the regulatory basics and of the financial benefits of appropriate cybersecurity, targeted at all actors, but especially senior management

- The proactive identification and coordination with other regulatory organizations, by aviation/airworthiness regulators (e.g., FAA, EASA) assisted by cross-sector knowledge hubs such as G-32; such coordination might require amendments to other sectors' regulation (e.g., GDPR) so the regulators are best equipped to lead this effort

- Enhanced education, in advance, perhaps even mandatory training of rookie applicants; such education can start with just the basic concepts (i.e., training for individual decision makers for such applicant) and then build up to full training as funds become available (e.g., for the first large contract including cybersecurity regulatory requirements)

- Immediate implementation, even voluntarily, of ARP4754/4761, preferably the upcoming ARP4754B and ARP4761A, by established actors that are already certifying aircraft and safety-critical systems, as these are embedded in the new airworthiness cybersecurity regulation, and established actors would be better off check-marking this task in advance

- Crafting the regulatory mandates so as to create a three-tier regulatory framework: high-level regulatory documents (i.e., IACO plus FAA and EASA), mid-level practical guides (i.e., DO-326/ED-202 set), and low-level technical best practices (e.g., JA7496, JA6678, JA6801, etc.)

- Initiating internal and external discussions by regulators, in order to work toward achieving the proper expertise balance with regulators, consultants, and applicants

# Current Standards/ Guidance/Best Practices Gaps

The current versions of the DO-326/ED-202 set are already complete and ready for prime time. In fact, the set is a sound foundation for the initial certification efforts expected in the early 2020s. However, in rushing to make it happen (yes, 15 years is considered to be "rushing" for regulation), certain important gaps remain in the set, that at first will need to be attended to on a case-by-case basis. As certification starts in earnest, these gaps would need to be treated in an orderly manner. Such major conceptual gaps are

- Insider threats
- Human factors
- Non-technology-oriented subjects
- Resilience
- Periodic activities
- Recertifying after being compromised

## Insider Threats

As is the case with cybersecurity models and standards elsewhere, one of the least attended to, yet one of the most abundant types of cyber-threats—with arguably more than 60% of all cyber-attacks related to it—is the insider threat [95, 96]. This, on top of being a potentially omni-threat (thus difficult to mitigate), also defies many current standard and model concepts. Indeed, the security architecture principles of DO-356A/ED-203A include some generic mitigations against unauthorized access, and the in-service phase of aircraft lifecycle represented by DO-355/ED-204 and the planned DO/ED-ISEM briefly touch authorization and access issues. Still, the conceptual background, which should lead to a coherent mitigation strategy and to solid certification, is simply not there.

For generic cybersecurity, there are already a few attempts at including this aspect into existing cybersecurity models, such as extending the Lockheed-Martin Cyber Kill Chain® [97] (Figure 19), which can serve as guidelines to the development of proper enhancements to the existing DO-326/ED-202 set.

**FIGURE 19.** **Lockheed-Martin Cyber Kill Chain®.**

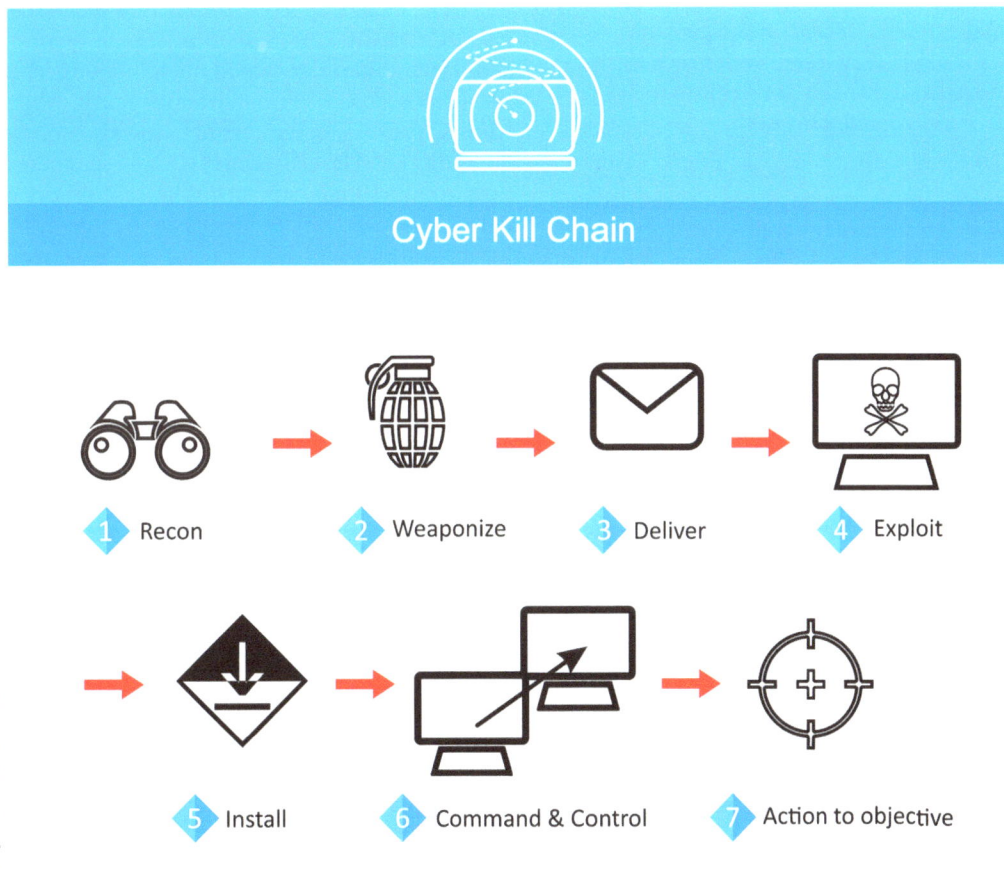

Maryna Yakovchuk/Shutterstock.com

# Human Factors

Although not the conceptual challenge that the "insider threat" poses, the human factor is still a crucial one in any and all mitigation techniques related to cybersecurity. Insider threats—and to an even greater extent, the human factor in general—are treated by DO-355/ED-204 and the planned DO/ED-ISEM. However, there are still at least two conceptual gaps at the regulatory level that need to be attended:

1. A conceptual gap, driven by the different, almost opposite, approaches to safety and security, whereas:

   a. The safety culture is already completely embedded in aviation, to the point that "whistleblowers" are treated as heroes rather than villains [98] and (at least in the US) one would usually not face criminal charges for disclosing own safety issues.

   b. The security culture is still very much a cover-up culture, partly because too much disclosure may expose the revealing party to greater cyber-threats. As this is mainly (but by no means solely) a cultural aspect, it is quite difficult to handle with regulatory tools, so the most plausible method might be through such channels as the Aviation Information Sharing and Analysis Center (A-ISAC) and the European Centre for Cybersecurity in Aviation (ECCSA), granting a certain level of confidentiality to members. However, either the A-ISAC/ECCSA scope needs to somewhat extend to include more personal aspects or the regulatory framework needs to adopt some A-ISAC/ECCSA aspects into the regulation "core"— preferably both.

2. A missing portion of the development phase, which will at least touch human aspects during development, potentially adopting some of the concepts discussed earlier for the conceptual gap. As the current DO-326/ED-202 set explicitly recognizes some of this gap (e.g., security training for all developers, not just for security experts), some progress should be expected for the next updates of the set.

# Non-Technology-Oriented Subjects

As the DO-326/ED-202 set is based on mainly aerospace industry technically oriented foundations, such as ARP4754/4761, DO-178/254, and so forth, the focus of the set is naturally technical. Consequentially, the core of the DO-326/ED-202 set, especially the parts concerning the development and assurance of the security architecture and measures development, heavily leans toward the development of software and somewhat hardware. Yet, it is not even remotely as descriptive when it comes to other aspects of cybersecurity, such as administrative measures, for instance.

This attitude is especially noticeable for the development phase, whereas the in-service phase guidance includes reasonable non-technology-oriented measures. Such an attitude served very well the purposes of certifying safety-critical systems for safety against "natural causes" (best described in technical terms) for decades—but not so much against human causes, even if technically delivered.

The enhanced inclusion of non-software/hardware in the development and assurance processes recommended for the DO-326/ED-202 set can be facilitated through either direct enhancements to the DO-326/ED-202 set or through SAE's G-32 in support of the DO-326/ED-202 set, and references to such supporting processes in the DO-326/ED-202 set. Sources to draw from are numerous—every common cybersecurity set of standards/best practices includes at least a minimal reference to such aspects—even DO-355/ED-204 of the DO-326/ED-202 set includes it, but only for the in-service phase, not yet for the development process.

# Resilience

Although, as previously discussed, the DO-326/ED-202 set is mostly technically oriented, there are still some technical-guidance topics missing—although present in the ARP4754/4761 set on which the DO-326/ED-202 set is based, at least at the assessment and system levels. The most notable such gap is arguably some form of redundancy: an instrumental feature of the ARP4754/4761 set.

It is, of course, a completely different issue for security, as any added features, such as simple redundant systems, similar or dissimilar, would obviously unnecessarily extend the attack surface of the aircraft/system, while (at least for similar systems, in the case of hardware) not even providing any additional level of security. However, there may be value to clear and simple explicit guidance, considerations, and/or encouragement for applicants to include in their architecture such means of added resilience as completely isolated backup systems, which are the actual security equivalent of safety's redundancy. A similar approach was applied by NSA's International Automotive Task Force (IATF) for recommended availability supporting mechanisms [99].

Such encouragement could be in the form of some credit to the applicant toward certification. This approach can be discussed either for updates to the DO-326/ED-202 set, or through SAE's G-32 in support of the DO-326/ED-202 set and references to such supporting processes in the DO-326/ED-202 set.

# Periodic Activities

The current in-service aspects of the DO-326/ED-202 set, mainly DO-355/ED-204, are arguably the most mature of the set, built on very solid foundations of similar previous standards, thus voluntarily implemented by quite a few actors in the aviation ecosystem even before the entire DO-326/ED-202 set becomes mandatory. However, some of the wide gaps in the set relate to the deployment phase (i.e., the transition from development and production to operations). The deployment phase can be seen as the formal handshake between the development parts of the set (i.e., DO-326A/ED-202A and DO-356A/ED-203A and the in-service parts: DO-355/ED-204 and its planned companion, DO/ED-ISEM).

Although the development parts of the set do specify such a handshake and do call for developers to provide the operators with proper means for continued airworthiness, the general attitude of the text is "event driven" rather than proactive, so that the operators and the developers would be expected to act upon the discovery of new threats, vulnerabilities, and so on. There is very limited mention of any proactive means, such as periodic risk analysis, penetration testing ("pen testing"), benchmarking, conformity reviews, or the like. There is no practical difficulty in adding such means, at least not the same activities that are already recommended for the development phase itself, in a periodic manner. It is advisable for the next updates of the DO-326/ED-202 set to include guidance for the developers to develop such explicit periodic activities' mandates: the most efficient way forward could be to adopt such practices from other ecosystems, such as the periodic activities mandated by the Commodity Futures Trading Commission (CFTC) [100].

## Recertifying After Being Compromised

Another aspect of continued airworthiness, which can also be related to the same deployment phase, is redeployment, or recertification following events and incidents, more so when cybersecurity is that aspect, and for the following key reasons:

- The mere number of cybersecurity events/incidents reports is at a completely different order-of-magnitude than "pure" safety events/incidents, even considering only the reports that are applicable to aviation CPSs.

- Cyber-attacks that might seriously challenge airworthiness would typically be carried out by sophisticated attackers, so unlike "pure" safety, investigations might be very challenging, and in more cases than for "pure" safety—the full picture may forever remain uncertain.

- The recertification process of a compromised system might in itself be challenging, as apart from the high number of expected events/incidents and the pressing need to resume safe flights as early as possible, there is a need for (at least) best practices for resuming a sufficient cybersecurity level, while trying to avoid, for as much as practical, a complete Risk Assessment process from scratch. A possible solution could be "certification/recertification institutions/labs" that would be certified for the task by the applicable regulatory authorities, along the principles of maintenance, repair, and overhaul (MRO) activities.

The current planned new DO/ED-ISEM (still no numbers allocated) "Guidance on Security Event Management" document, scheduled within a year or so, might be a step in that direction as a process document if this gap is to be handled under the scope of the first version of the document. Still, best practices would need to be developed to support it—or any future document that would close this gap. This gap and recommendation to close it are in line with the AIA 2019 cybersecurity report [101].

## Recommendations

Major gaps in the DO-326/ED-202 set guidance discussed in this report include missing or lacking reference to

- Insider threats

- Human factors in general

- Issues and solutions that are non-technological

- Resilience-improving recommendations

- Mandates for periodic detective/preventive activities

- Practices for recertification after being compromised

It is advised that all aviation cybersecurity forums and committees dedicated to aviation cybersecurity guidance—including technical, industry, and regulation aspects at SAE, RTCA, EUROCAE, FAA, EASA, and the like—look into these gaps and provide suggested solutions that can later be discussed and adopted.

## Current Standards/ Guidance/Best Practices Inherent Dilemmas

Unlike gaps that may be simply closed with proper guidance from standard makers, this unsettled domain touches the inherent dilemmas of the DO-326/ED-202 set: contradicting guidance or conceptual realities that cannot be easily settled. Some may not be settled at all in their current form and may require paradigm shifts in order to be settled. Some dilemmas are explicitly acknowledged by the text of the DO-326/ED-202 set, while others require deeper understanding of the set. Such major inherent dilemmas are

- Simplicity versus complexity

- "Work in Progress"

- Binary trust versus standard segmentation

- The ever-evolving threats

- "Slow" safety versus "Fast" security

- Which Risk Assessment approach to use?

## Simplicity Versus Complexity

A key principle asserted by the DO-326/ED-202 set (for the proposed Security Architecture) is "Keep It Simple," and in fact, it makes a lot of sense even beyond that at the regional and global regulatory levels. Airworthiness cybersecurity is inherently complex, so complicating its regulatory framework might further delay it and/or lower its effectiveness.

However, as the aviation sector is so complex, and regulatory frameworks and techniques are developed by separate organizations, not all of which are even aware of each other's efforts, the current regulatory landscape is extremely convoluted, fractured, and at times, even incoherent across nations, sectors, actors, and disciplines. Even the indispensable DO-326/ED-202 set took almost 15 years to produce a complete, useful set which is still quite fractured, definitely imperfect, and might require many more years to settle.

This inherent dilemma is very hard to reconcile and would require the combined efforts of all relevant organizations to coordinate the multiple moving parts of regulation, and these parts are still in motion. As a consequence, this dilemma would most likely not be resolved, for at least the current decade.

## "Work in Progress"

One of the most candid statements in the DO-326/ED-202 set asserts that the document, set, regulation, and entire methodology are still very much a "Work in Progress." This state of things is quite unprecedented in aviation regulation—as although many aviation practices and standards keep evolving, the frameworks tend to eventually stabilize following an initial period of settling down. In contrast, even the frameworks of cybersecurity may continually evolve, because threats continuously evolve, and we have a dynamic double-edged sword of change.

The practical implication of such a statement (that is very much correct) is that no applicant or regulator can be completely certain as to the regulatory process or even the framework—let alone the fine details of the certification process. The dilemma here is clear: whether to have a "temporarily stable" set of regulations, but go ahead and change it every few years, turning parts of the industry upside-down; or else have some sort of a dynamic set of rules that will constantly keep the industry out of its comfort zone.

A creative way of settling this dilemma could be for the regulation to explicitly allow for processes/requirements to evolve, along the lines of the National Initiative for Cybersecurity Education (NICE)/Comprehensive National Cybersecurity Initiative's (CNCI's) Cybersecurity Capability Maturity Model (CCMM/C2M2) [102], the Department of Defense's Capability Maturity Model for Cybersecurity CMMC [103] (Figure 21), Lockheed-Martin's Cyber Resiliency Level™ (CRL™) [104], or any other cybersecurity adaptations of the classic Capability Maturity Model Integration (CMMI)

[105] model (Figure 20). Such an approach, also promoted by other prominent organizations, like AIA [106], Open Web Application Security Project (OWASP), Software Assurance Maturity Model (SAMM), and others, when adapted from confidentiality oriented models such as the ones mentioned above to an aviation safety-critical approach, can mitigate the effects of an always-changing environment and even allow for yet stricter levels of cybersecurity to be included in the future.

## Binary Trust Versus Standard Segmentation

"Trust" is one of the most important attributes concerning cybersecurity, and as such, assumes a prime role in aviation cybersecurity regulation. The 2018 updates of DO-356/ED-203 to revision-A saw the transition from a "five-shades-of-grey trustworthiness scale" (DO-356) or no scale at all (ED-203)—to a binary, black-and-white quality (whereas according to the original DO-356, "Trust" was a quantity measured on a five-level scale). Now, with DO-356A/ED-203A, any element in the security environment can only be defined as "Trustworthy" or "Untrustworthy," so the security perimeter can now become crystal clear.

However, although only in an appendix, rather than in the body of the document, the new DO-356A/ED-203A now include a strong recommendation in favor of segmentation, by applying the ARINC 664-5 [107]/ARINC 811 [108] domains approach (Figure 22), which is the de facto best practice of the aviation industry since even before the DO-326/ED-202 set. The domain approach, as applied to commercial aircraft, calls for four domains—assigned gradual trust levels—from the most trustworthy domain at the core of the aircraft systems, to a completely untrustworthy domain outside the aircraft's hard-wired systems.

These two approaches, each very instrumental to airworthiness cybersecurity, cannot reside together in their current form. This dilemma calls for deep conceptual discussions in order to be resolved, maybe even resulting in conceptual changes for one of them or both. A practical approach toward resolving it could be: keeping the domain approach in future DO-356/ED-203 versions at a top-level form (even if it is to remain an appendix) while reconciling it with the binary trust approach in deeper, more technically oriented organizations such as SAE International through G-32 and/or SAE-ITC (ARINC) itself.

## The Ever-Evolving Threats

Classic safety events rarely pose new types of failures or errors—as, by definition, none is a deliberate act. Cyber-attacks, on the other hand, may evolve to pose new types of threats quite often ("often" on a regulatory time scale). Furthermore, the rate at which new threat types are discovered, let alone developed, is strikingly high.

**FIGURE 20.** Original Capability Maturity Model Integration (CMMI®).

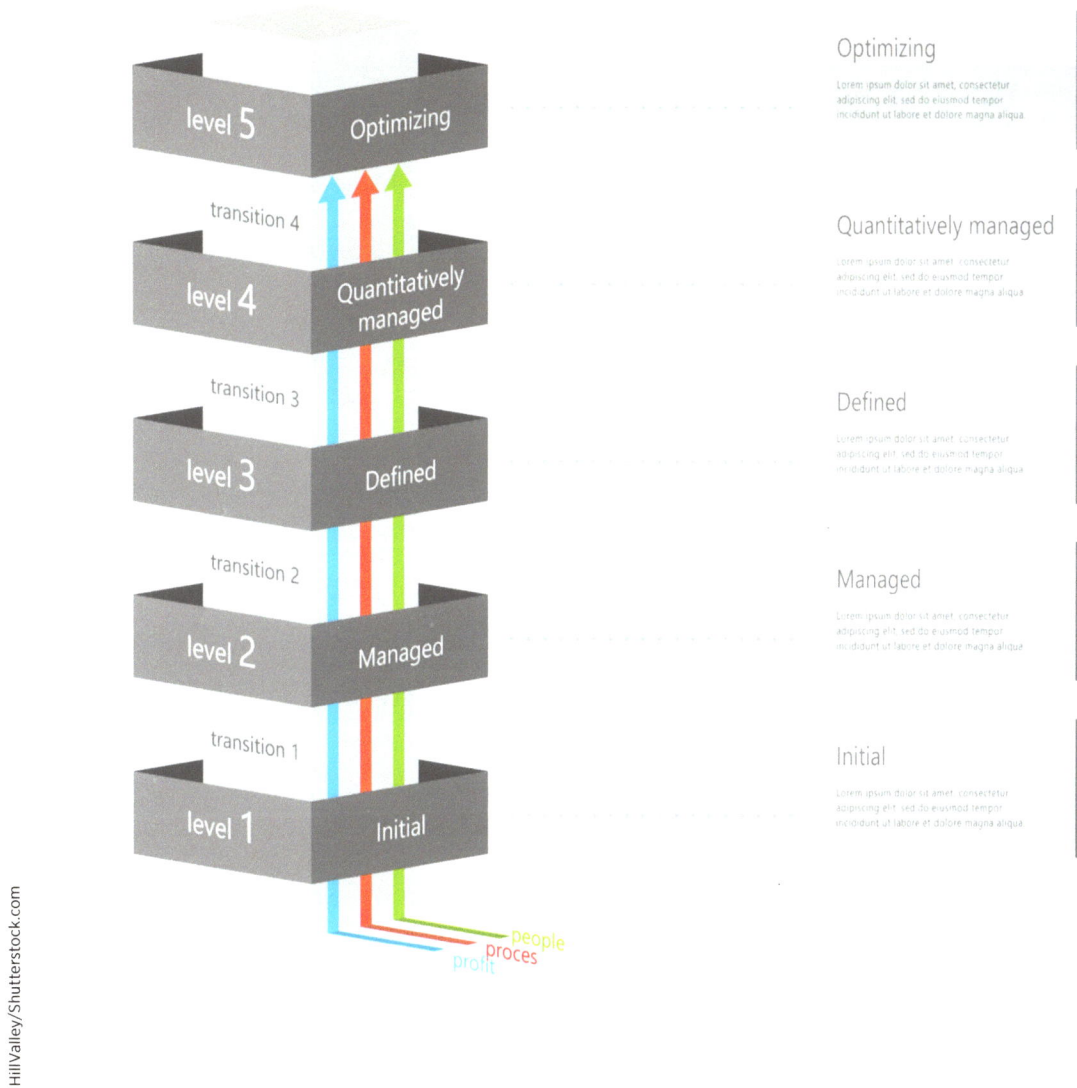

**FIGURE 21.** The Department of Defense's Capability Maturity Model for Cybersecurity (CMMC).

**FIGURE 22.** Aircraft Cybersecurity Domains.

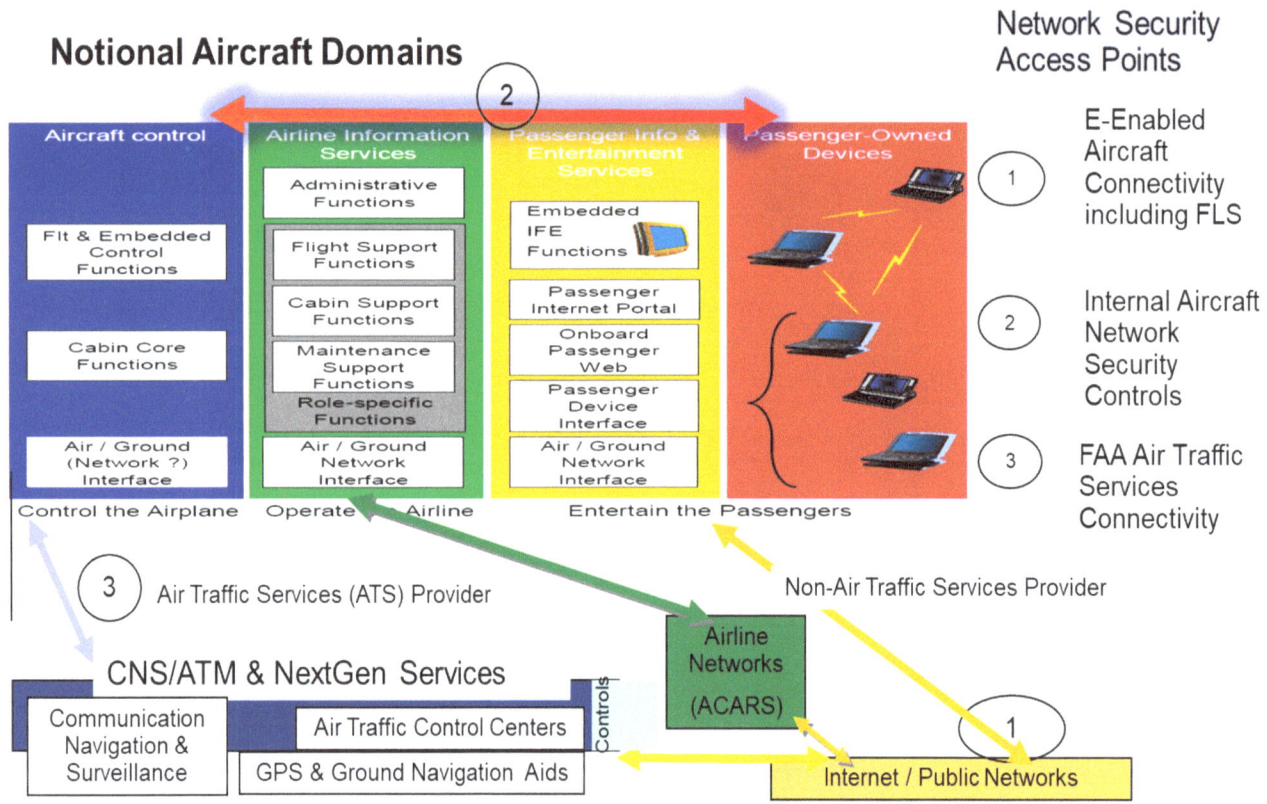

This is a well-known cybersecurity issue, and the current DO-326/ED-202 set keeps relating to it, to the point that it almost forces the conclusion that there is nothing really effective that could be done about it, except for chasing the constantly evolving threats, weaknesses, and vulnerabilities (periodically or per event).

Although this is a dilemma unlikely to go away, it may still be helpful to explore some more proactive methods, such as the moving target defense [109, 110], active defense [111], and/or other techniques, to improve the odds. Such techniques might be still immature; however, the aviation sector, with safety requirements way beyond other sectors, can encourage applicable research and development to make such techniques available later in the decade in order to progress faster than potential adversaries. The DO-326/ED-202 set might just allow for such proactive approaches in principle, while the specific methods would be best served by being dealt with under the auspices of technical forums such as SAE G-32.

## "Slow" Safety Versus "Fast" Security

The rapid pace of cyber-threat emergence, forcing a rapid cybersecurity pace for solutions to counter these threats, is the source for yet another dilemma beyond cybersecurity unique issues. In comparison to cybersecurity's naturally fast pace, with software updates over the air (SOTA) as a mainstream process for applying quick modifications (Figure 23). Safety, on the other hand, is, by nature, a very slow discipline, as it requires exhaustive, deep, and lengthy processes. The odds are simply too high when dealing with airworthiness, so no shortcuts can be allowed.

The introduction of cybersecurity alongside safety is destined to present new types of friction between these two completely time-wise opposite disciplines, which would present regulators and applicants alike with yet-unknown dilemmas: Long/short development cycles? Slow/fast updates? Who should own the combined process, be it certification, assessment, development assurance, or anything else—safety personnel or security personnel ("slow" or "fast" lead process)? What level of integration to apply to any of the processes, and what level to the entire process: tightly coupled, completely separated except for predesignated reviews, anywhere in between? Such decisions are not merely procedural, as they are certain to affect the outcome, so the stakes for this dilemma are high.

In the early years of DO-326/ED-202 set development, an all-inclusive "safety-first-and-last" approach took over regarding security as a simple subset of safety; while currently, in many cases, experts are granted broad authority over the

**FIGURE 23.** Software updates over the air (SOTA)-an automotive example.

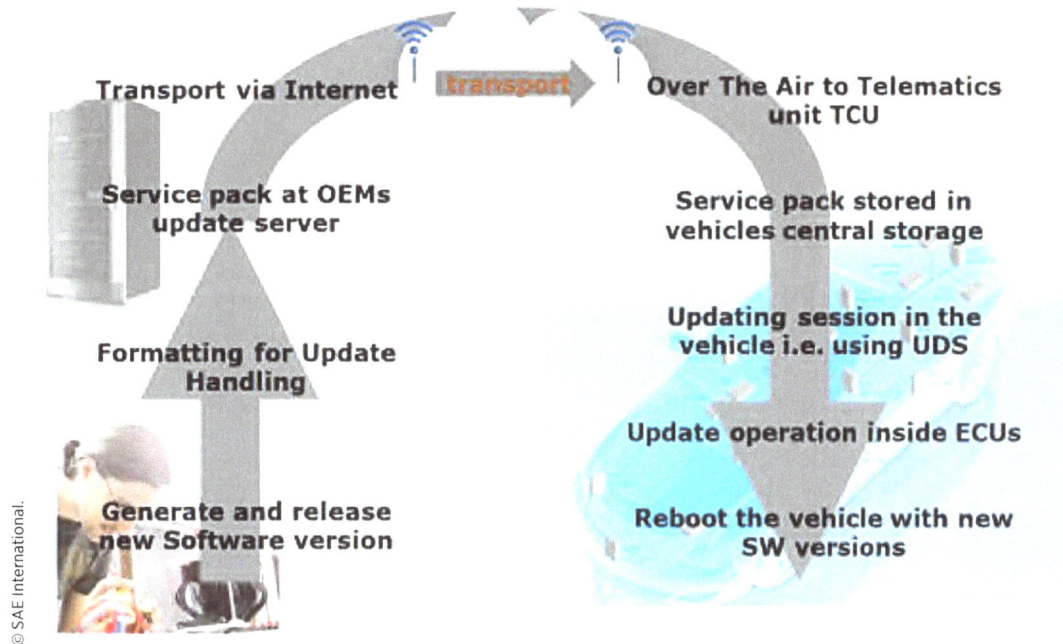

entire process, not always realizing the deep safety aspects of airworthiness: almost the opposite approach. Neither extreme seems to be the solution. This dilemma is here to stay, and any resolutions should probably be carefully crafted middle-of-the-road-type solutions, with very broad flexibility to be applied based on each specific case details. Even the decisions of which of these dilemmas to handle under what organization—and which should remain open—needs to be carefully addressed. This calls for joint action by regulators, guidance developers (e.g., RTCA, EUROCAE), and standard makers (e.g., SAE), and is not expected to be an ad hoc activity or even series of activities, but an ongoing effort.

## Which Risk Assessment Approach to Use?

Among the major dilemmas related to the DO-326/ED-202 set, there is one that captured the headlines of the aviation cybersecurity regulation community for a few years during the 2010s: the proper approach to risk assessment, one of the cornerstones of airworthiness cybersecurity certification. In 2014, RTCA released DO-356, using the "Likelihood Method" Strategy, and in 2015 EUROCAE released ED-203, using the "Effectiveness Method" strategy—different approaches for the same process (Figure 24).

Although theoretically both methods should yield the same bottom-line results, the mere fact that the way to reach the results was not the same would have, in practice, yielded different results. Worse yet, as one method was American and the other

European, applicants would have faced uncharted waters in the certification process, as no one could be certain of being certified for both unless performing two risk assessments, using both methods. One of the most pressing issues on the desk of the FAA ARAC ASISP WG [112] (active 2014 to 2016) was the harmonization of DO-356 and ED-203, differing on other subjects as well, and the result was a clear WG recommendation to somehow update the two documents so they become technically identical.

The way DO-356 and ED-203 were finally harmonized was a typical compromise process: neither method "won" and the subject was left open, downgrading both methods to become separate appendices in the new DO-356A/ED-203A harmonized documents, but it also added two new optional methods for the same process: System Theoretic Process Analysis-Security (STPA-Sec) and the Cyber Risk Assessment Model (CRAM). The top-level definition was unified and simply named "Level of Threat" (Figure 25). So now, instead of having *two* unsettled options, applicants have *four* unsettled options to choose from, and the dilemma just doubled (at least). That was the price of achieving the mandate of mutual acceptance of all methods by both US and EU regulators and reaching a complete DO-326/ED-202 set.

Obviously, as this compromise (not the best term when cybersecurity is at stake) is basically political, it would be extremely difficult to resolve this dilemma. However, at the technical level, in the likes of G-32, it may be advisable to develop, over time, best practices to help applicants to better approach the problem, perhaps even develop a meta methodology for selecting the proper method.

Parting shot at this subject: As the efforts to harmonize ED-201 with an RTCA matching DO are still in progress,

**FIGURE 24.** Level of Threat: Original DO-356/ED-203.

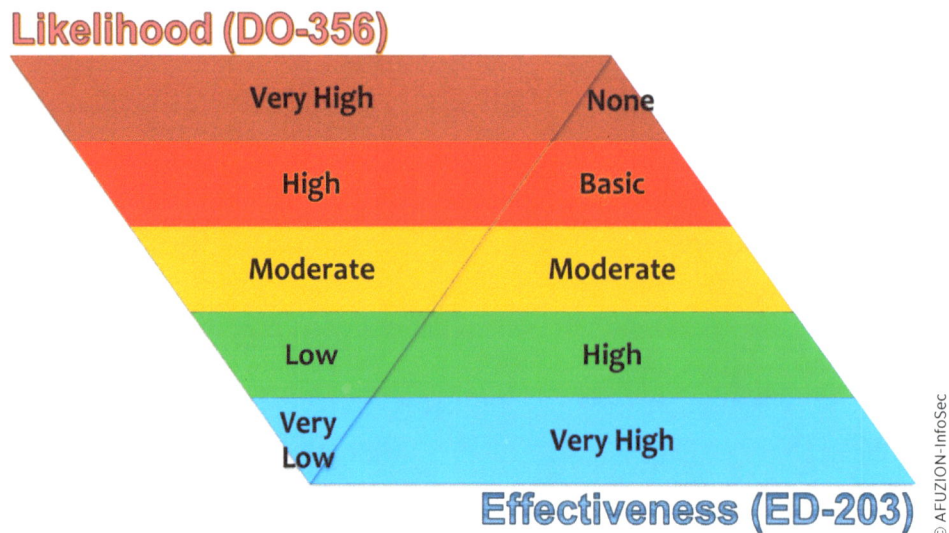

**FIGURE 25.** Level of Threat: Unified DO-356A/ED-203A.

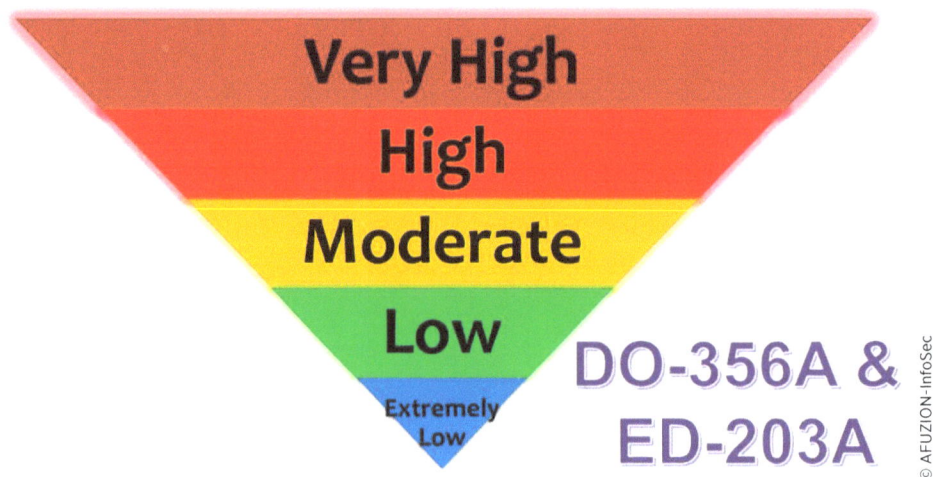

regulators and applicants might need to follow closely, in case these two much more strategic documents drift apart, to avoid even worse headaches for airworthiness cybersecurity regulation.

# Recommendations

Unlike previously specified DO-326/ED-202 set gaps, the dilemmas covered by this section are inherent in the set, and quite a few of them are even explicitly recognized by the set authors in the text. The dilemmas identified here were:

- How simple or complex should the regulation be?

- How to treat this eternally evolving regulation, as by its nature, cybersecurity regulation cannot be realistically settled?

- How to accommodate the black/white "trustworthiness" attribute with the adoption of the practical and popular "aircraft domains approach," which does imply more than two possible levels of trustworthiness?

- How to advance from current classic relatively passive cybersecurity approaches prevalent throughout the DO-326/ED-202 set to more proactive techniques?

- How to accommodate safety, an inherently "slow" discipline, with security, an inherently "fast" discipline?

- How to select the proper risk assessment methodology out of the four options currently recognized by DO-356A/ED-202A?

It is advised that all aviation cybersecurity forums and committees—especially technical and industry aspects at SAE, RTCA, and EUROCAE—look into these dilemmas, analyze them thoroughly (maybe with academia), and provide suggested solutions that can later be discussed and adopted.

# Current Standards/ Guidance/Best Practices Methodology Uncertainties

All the previous unsettled domains dealt with direct practical aspects of the DO-326/ED-202 set, but this final domain poses unsettling questions regarding some of the set's current methodologies themselves. Are we even heading in the right direction? Even if so, how likely is this direction to avoid methodological dead ends going forward? Such methodological uncertainties concerning the DO-326/ED-202 set are

- Process or requirements?
- An open-ended methodology?
- "Enumerating Badness?"
- What are the REAL odds?
- Federated no more?
- The future looms large

## Process or Requirements?

The overall attitude of the DO-326/ED-202 set mandates is process-oriented, very much like the primary sources from which it draws: ARP4754/4761, DO-178/254, and the like. However, there is one fundamental difference between these safety-oriented sources and the security-oriented DO-326/ED-202 set: whereas the classic ARP4754/4761, DO-178/254 are mostly about assuring quality and other, mainly passive objectives to enhance safety, the DO-326/ED-202 set does need at least some minimal requirements/specifications for security, even if not for specific security measures.

Furthermore, even for the classic ARP4754/4761, DO-178/254 sets, as well as safety-critical applications in general, wrong or subpar requirements are a major cause for defects [113].

In other words, the process-only passive methodology itself breaks down when applied to cybersecurity, as securing against cyber-threats requires specific standards for actions, dedicated hardware, and/or software and more. Such specific requirements are only vaguely addressed in the DO-326/ED-202 set, so that minimal compliance, to avoid high cost, may lead to inappropriate and/or inadequate solution requirements. In fact, the examples most adequate for a discipline like cybersecurity might actually be, at least in part, the DO and ED documents for "Minimum Performance"/"Minimum Requirements," such as for radios, radars, and the like. Even cybersecurity guidance dating back a few decades, such as NSA's IATF, includes requirements for strength of mechanism levels (SML) alongside evaluation assurance levels (EAL) [114], adjusted to asset value and threat severity. EAL, originally developed within the Common Criteria (CC) [115], could be seen as the parallel of the DO-326/ED-202 set's SAL, but SML has no current aviation regulatory parallel, not to mention IATF's distinction of various types of measures for various types of security attributes (confidentiality, availability, etc.), which can be considered as a third dimension of requirements.

The implication is, that either the DO-326/ED-202 set should include clear performance/requirements-oriented parts, making processes—as crucial as they are—only a companion of the requirements parts. That, or introduce a clear distinction inside the set between process-oriented and performance/requirements-oriented documents, maybe even leaving the "Minimum Requirements" for technical standard-making committees such as SAE G-32. Such Minimum Requirements can be inspired by IATF's elaborated SMLs, coupled with IATF's/CC's EALs [116], to stay in line with other common cybersecurity standards. One immediate option that can already be promoted is the usage of blockchain technology [117], as the vehicle for such a standard-making effort already exists at SAE: committee G-31 (that can be joined by G-32 for one potentially powerful functional solution, in addition to other potential solutions) (Table 1).

Isolating and enhancing the process parts of the DO-326/ED-202 set is deep within the mandates of both WG-72 (EUROCAE) and SC-216, while the performance/requirements parts can be dealt with by G-32 or jointly with WG-72 and SC-216.

**TABLE 1.** IATF p.4-32, Table 4-7, Degree of Robustness.

| Information Value | Threat Levels | | | | | | |
|---|---|---|---|---|---|---|---|
| | T1 | T2 | T3 | T4 | T5 | T6 | T7 |
| V1 | SML1 EAL1 | SML1 EAL1 | SML1 EAL1 | SML1 EAL2 | SML1 EAL2 | SML1 EAL2 | SML1 EAL2 |
| V2 | SML1 EAL1 | SML1 EAL1 | SML1 EAL1 | SML2 EAL2 | SML2 EAL2 | SML2 EAL3 | SML2 EAL3 |
| V3 | SML1 EAL1 | SML1 EAL2 | SML1 EAL2 | SML2 EAL3 | SML2 EAL3 | SML2 EAL4 | SML2 EAL4 |
| V4 | SML2 EAL1 | SML2 EAL2 | SML2 EAL3 | SML3 EAL4 | SML3 EAL5 | SML3 EAL5 | SML3 EAL6 |
| V5 | SML2 EAL2 | SML2 EAL3 | SML3 EAL4 | SML3 EAL5 | SML3 EAL6 | SML3 EAL6 | SML3 EAL7 |

Reprinted from NSA document. Defense Technical Information Center. U.S. Government Work

# An Open-Ended Methodology?

When addressing the procedural security assurance attributes of the different security measures within the DO-326/ED-202 set, one of the most important processes of the set—an underlying methodology unfolding almost as an afterthought—is assigning security measures per threat scenarios, rather than per "Things" (e.g., software/hardware elements, as in the "Internet of Things") as the classic ARP4754/4761, DO-178/254 sets do, when transitioning from safety assessment to development assurance. Although DO-356A/ED-203A still allow assigning a specific security measure per multiple threat scenarios, this approach is a completely open-ended requirement, that has the potential of keeping the entire design from ever converging, especially having in mind the ever-evolving nature of cybersecurity threats.

The tacit determinism killer is "change over time:" although both safety hazard-risk assessments and security threat-risk assessments are dynamic and can produce a large number of cases/scenarios, safety assessments "freeze" upon transition to the design phase, whereas security assessments keep "festering" as new attack methods are developed, common vulnerabilities and exposures (CVEs [118]) are reported faster, and new weaknesses emerge even faster than that. Whereas in 1992, the year DO-178B was published, approximately one new CVE was reported per month, in 2006 to 2007, the years WG-72 and SC-261 were launched, this rate was already more than one CVE per 90 minutes, and as of December 2019, this rate was already one per 23 minutes—these are not viable rates for robust engineering design. The rapid rate of new vulnerabilities can be clearly realized from Figure 26, describing the accumulation of reported CVEs since such reports commenced in 1988 through 2019.

Hence, safety can get to a reasonable closure, whereas security probably cannot.

Although threat assessment is vital for any cybersecurity plan and activity to be valid, this open-ended methodology

**FIGURE 26.**   Cumulative CVEs per year, 1988-2019, calculated from NIST NVD, January 2020.

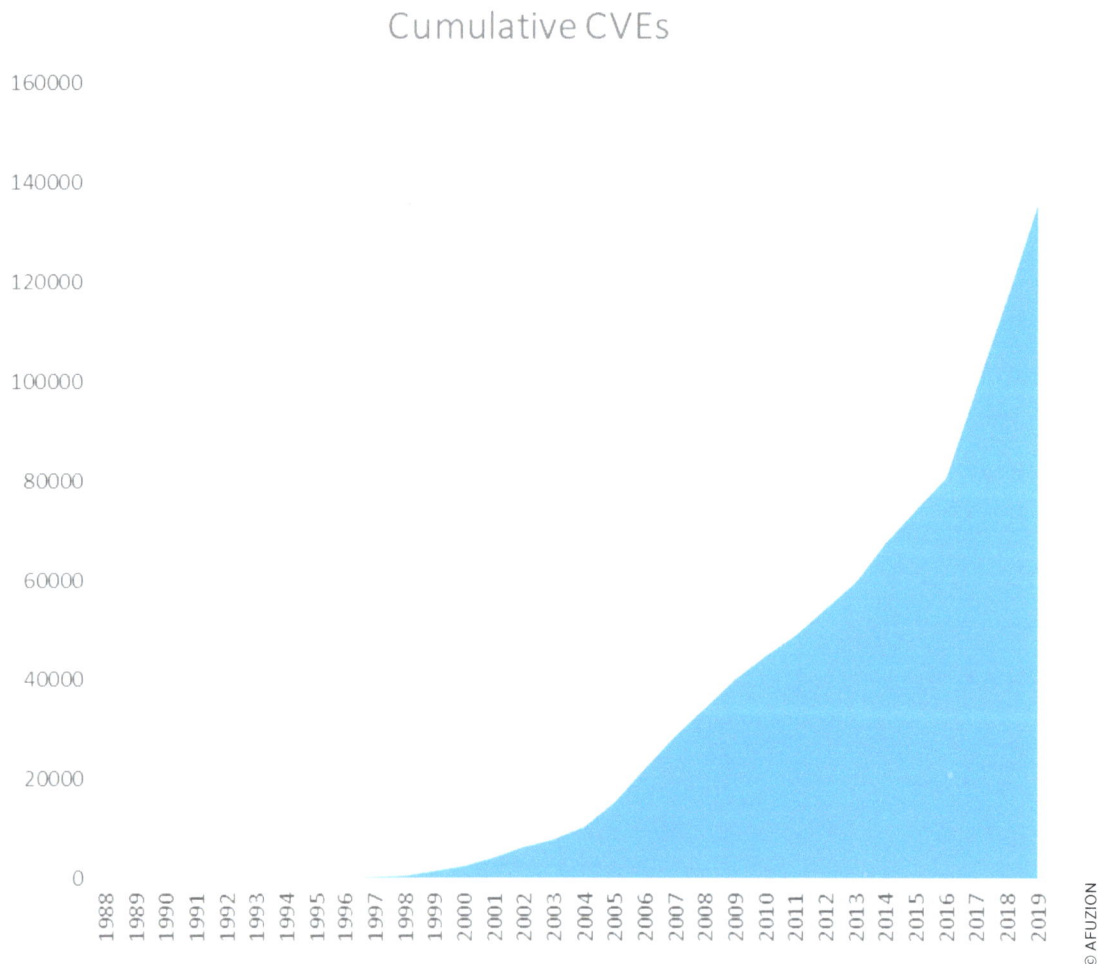

of requirements setting is prone to producing shortcuts on one hand and overkills on the other hand, depending on the stakeholders' positions and resources. A more viable methodology could be based on "Things," using threat scenarios just as a (very useful) tool for specifying security architecture and requirements and/or as a (very powerful) tool for validation and verification, for instance, but decoupled from the actual design at a certain stage of the development process.

As this methodology touches mandatory requirements, its updates, in any way, shape, or form need to flow through WG-72/SC-216 and FAA/EASA; however, forums like G-32 can heavily support the technical underpinnings of such updates.

## "Enumerating Badness?"

Further zooming in on the threat-oriented methodology, an even deeper methodological concern might be "Enumerating Badness" [119] (i.e., emphasizing external threats over emphasizing internal solid engineering and best practices). Indeed, the DO-326/ED-202 set does call for hardening software and hardware, but even this is just for technical security means, and as a response to threats based on threat scenarios. Furthermore, the DO-326/ED-202 set encourages the subscription to such resources as NIST's CVE, Common Weakness Enumeration (CWE) [120], Common Vulnerability Scoring System (CVSS) [121], and Common Attack Pattern Enumeration and Classification (CAPEC) [122], among others—all highly professional and top-quality resources, but still "Enumerating Badness."

The OWASP basic methodology points out that there is considerably more "Badness" than "Goodness," and that "Badness" cannot be controlled by defenders while growing at explosive rates. Attempting to "Enumerate Badness" is a futile exercise, so a better approach would be to focus on the existing or planned "Goodness" in a proactive way.

One potential outcome of stretching the "Badness" focus was shown in the previous aspect—arriving at an open-ended, never-converging process, and an option to alleviate the constant chase after such "Badness" was demonstrated for a previous unsettled domain—becoming proactive with such techniques as the moving target or active defense approach. Dealing with "Badness" should still occupy a great deal of any certification process, but unlike in the current mandates, probably not to a larger extent than faults/failures occupy for safety.

However, in order to emerge out of this existing state-of-mind, it would probably be advisable to cultivate technical solutions that can become part of the DO-326/ED-202 set and other regulation, through SAE G-32 in search for a more balanced approach. A good starting point could be a table-top review of existing techniques for safety hardening, such as the built-in test (BIT) [123], to mention but one: what could become the cybersecurity equivalent of initialization-BIT (I-BIT) and continuous-BIT (C-BIT), to considerably improve the aircraft's resilience? The security-BIT technique

was apparently already proposed as a mitigation technique by (at least) the FAA, but was eventually left out of the DO-326/ED-202 set, while practices of this nature are already gaining traction for Real-Time Embedded Safety Systems [124].

## What Are the REAL Odds?

How solid is the current risk-based methodology for cybersecurity in any event? This question may look somewhat out of context in such a discussion, but even the deepest and most obvious assumptions should be fairly examined.

When safety is at stake, statistical considerations have a lot of merit when coupled with severity in order to produce the overall risk of a hazard, as per ARP4761 (Figure 27). However, when designing software, the probability of a software defect is a priori assumed to be 100% certain [125]. The DO-326/ED-202 set itself acknowledges that information security failures do not have an expected failure rate, so for cybersecurity risk assessments it defines "Level of Threat," which according to the technique being applied, can be the likelihood of the threat materializing, or the effectiveness of the security measure(s).

However, apart from the unique aspect of the ever-evolving threat, and the a priori reliance on the existing or planned security measures (a recursive technique in itself), it is not at all clear that a threat, which would almost always materialize at least partly through software, and manipulated by humans to constantly seek new attack vectors and paths, should be even assigned any "likelihood" other than 100% or an "effectiveness" other than zero (for the long run). Remembering that the service life of aircraft is measured in decades, the long run is all but certain of arriving, sooner or later—rather sooner for cyber-threats.

This methodological uncertainty calls for a true open discussion, which may eventually result in collapsing the two-dimensional risk tables into a single-dimension risk as a function of only the severity of the outcome, or any other outcome that may challenge and/or update the classic safety refinement Function-Design-Architecture model (Figure 28).

## Federated No More?

As cyber-threats advance along attack paths, and could be considered to be percolating across the compromised system, a troubling general scenario may arise: could it be that independence assumptions, that are one of the risk-theory cornerstones, are refuted when assessing cyber-threats? And if so, would the implications include the "upgrade" of all severity levels of the analyzed element to the highest severity of the scenarios analyzed?

If this would be the case, then any analyzed system that cannot be proven to be completely isolated from cyber-threats would just become part of one large supersystem, which would be required to meet the highest level of assurance assigned to any single system of this supersystem. It can be argued that from a safety point of view, independence

**FIGURE 27.**    **ARP4761 Failure Condition Severity as Related to Probability Objectives and Assurance Levels.**

| Probability (Quantitative) | | Per flight hour | | | | |
|---|---|---|---|---|---|---|
| | | 1.0          1.0E-3 | | 1.0E-5 | 1.0E-7 | 1.0E-9 |
| Probability (Descriptive) | FAA | Probable | | Improbable | | Extremely Improbable |
| | JAA | Frequent | Reasonably Probable | Remote | Extremely Remote | Extremely Improbable |
| Failure Condition Severity Classification | FAA | Minor | | Major | Severe Major | Catastrophic |
| | JAA | Minor | | Major | Hazardous | Catastrophic |
| Failure Condition Effect | FAA & JAA | - slight reduction in safety margins<br>- slight increase in crew workload<br>- some inconvenience to occupants | | - significant reduction in safety margins or functional capabilities<br>- significant increase in crew workload or in conditions impairing crew efficiency<br>- some discomfort to occupants | - large reduction in safety margins or functional capabilities<br>- higher workload or physical distress such that the crew could not be relied upon to perform tasks accurately or completely<br>- adverse effects upon occupants | - all failure conditions which prevent continued safe flight and landing |
| Development Assurance Level | ARP 4754 | Level D | | Level C | Level B | Level A |

Note: A "No Safety Effect" Development Assurance Level E exists which may span any probability range.

© SAE International

can still remain very much intact, as most of the boundary breaching is due to deliberate penetration, and the classic AC/AMJ 25.1309-1 (B-Arsenal Draft) [126] excludes "sabotage" (e.g., cyber-attacks), hence, there is a solid rationale for the separation of different severity levels (Figure 29) and for separate DAL allocation. However, even if the implications are to be confined to security aspects only and "percolate" to the highest severity level within every system, these enhanced security approval requirements would still include the large portions relying on ARP4754/4761, DO-178/254, and so forth at the higher DALs, as these are part of the higher SALs of the DO-326/ED-202 set, which may at least partly do away with severity-level separation (Figure 30). Furthermore, the DO-326/ED-202 set, as previously demonstrated, tends to be more open ended with requirements. Additionally, when ascending the SAL ladder, a higher number security measures could be required, as is presented in Table 2.

This "dis-federation" avionic trend is not new, and there are even mature certification means [127] for it, preceding to cybersecurity mandates; however, the security effects described here might accelerate this trend beyond the current, relatively controlled, pace. Furthermore; the emerging risk-sharing concepts for the next revisions of ED-201 and its RTCA counterpart may accelerate this trend to include related ground systems as well, which may yet evolve into an issue of grand proportions.

This methodological uncertainty also calls for an open discussion, which may eventually result in modifications to the threat assessment models and/or assurance levels definitions for security, but maybe even have an effect on independence assumptions for safety assessments (thus on some of the most fundamental assumptions for federated avionics in general).

# The Future Looms Large

Perhaps the most intriguing uncertainties for current cybersecurity in general may yet emerge from next-generation technologies, which may challenge some of the most basic assumptions of today's common practices, techniques, or even methodologies. This report would touch only two of the most obvious such technologies.

An obvious threat to current cybersecurity assumptions is the emerging field of artificial intelligence (AI), which may shake, for instance, the very basic definition of a threat source. Whereas currently, there needs to be some sort of mal-intention by any part of the attack source (i.e., at least one of the links of an attack chain has to be human) artificial neural networks (ANN) can be developed by completely innocent humans, but develop their own, unintentional malicious agenda. Whom should the "mal-intention" aspect of a cyber-attack be related to in such a case? Naturally, ANN and other AI techniques can also present cybersecurity with an abundance of helpful security measures, but at least some of these measures may dramatically vary from current definitions of security measures. Aviation cybersecurity standardization and regulation efforts of all kinds would be advised to keep a close look at all times at their AI colleagues. For instance, SAE G-32 is advised to closely watch and update regarding

**FIGURE 28.** Overview of Development Process.

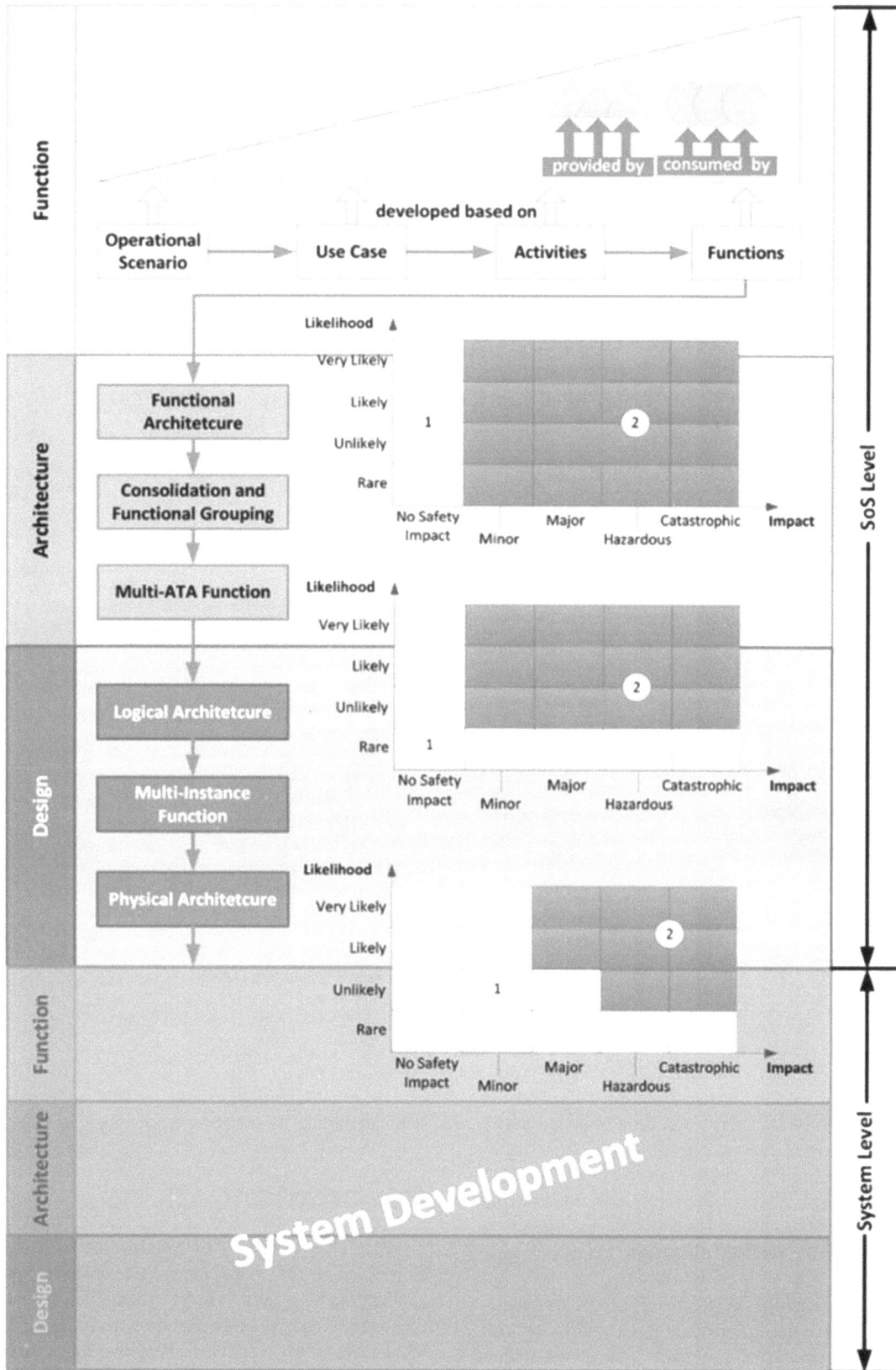

**FIGURE 29.**  **Examples of Classic Safety Severity Levels.**

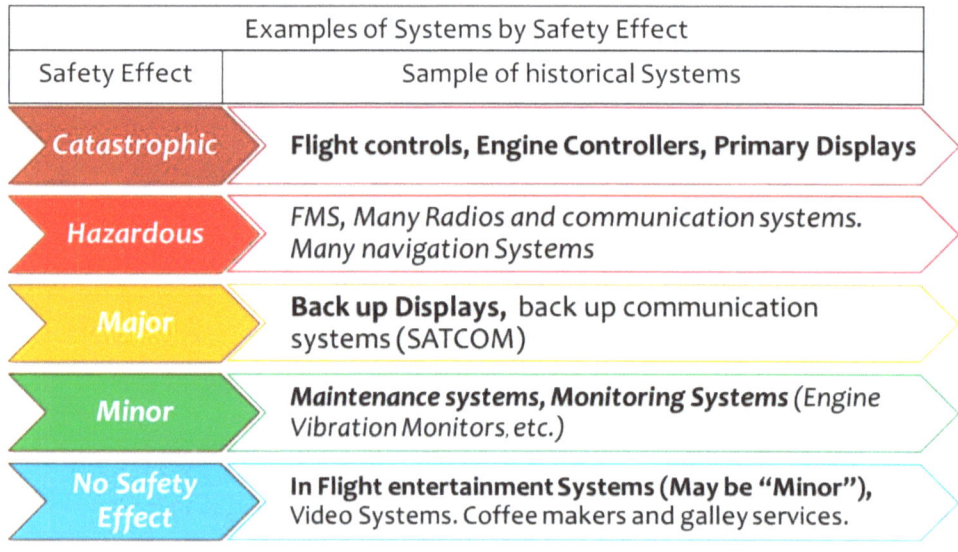

| Examples of Systems by Safety Effect | |
| --- | --- |
| Safety Effect | Sample of historical Systems |
| **Catastrophic** | **Flight controls, Engine Controllers, Primary Displays** |
| **Hazardous** | *FMS, Many Radios and communication systems. Many navigation Systems* |
| **Major** | **Back up Displays,** back up communication systems (SATCOM) |
| **Minor** | *Maintenance systems, Monitoring Systems (Engine Vibration Monitors, etc.)* |
| **No Safety Effect** | **In Flight entertainment Systems (May be "Minor"),** Video Systems. Coffee makers and galley services. |

© AFUZION-InfoSec

**FIGURE 30.**  **Potential Cybersecurity effect on Safety Severity Levels?**

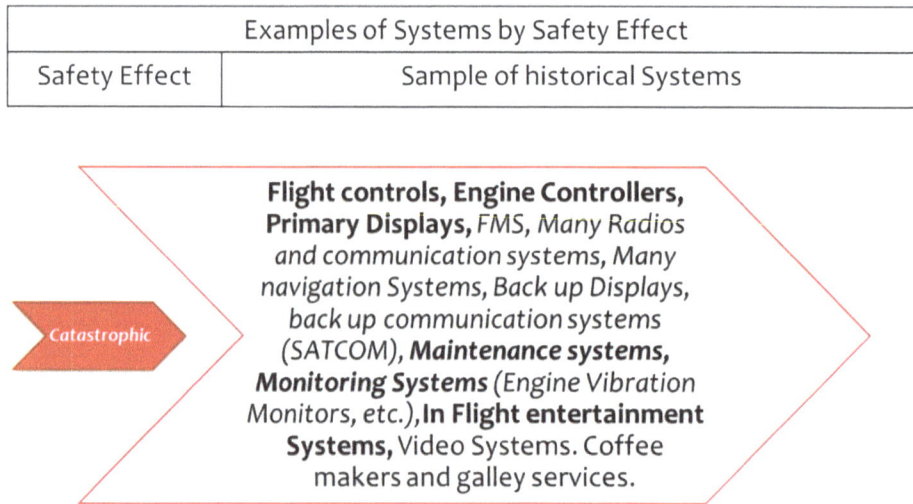

| Examples of Systems by Safety Effect | |
| --- | --- |
| Safety Effect | Sample of historical Systems |
| **Catastrophic** | **Flight controls, Engine Controllers, Primary Displays,** *FMS, Many Radios and communication systems, Many navigation Systems, Back up Displays, back up communication systems (SATCOM),* **Maintenance systems, Monitoring Systems** *(Engine Vibration Monitors, etc.),* **In Flight entertainment Systems,** Video Systems. Coffee makers and galley services. |

© AFUZION-InfoSec

SAE G-34 "Artificial Intelligence in Aviation." This report has made clear that cybersecurity guidelines required many years to mature; likewise, the emerging AI guidelines are in their infancy and will probably take years to evolve, in no small part due to the parallel consideration and evolution of cybersecurity. Nonetheless, the cross-committee cooperation designated here is an immediate mitigation that can already be applied.

Another emerging challenge to existing cybersecurity is quantum computing, with (at least) the potential of rendering many encryption techniques useless, which would probably be just the tip of the iceberg. Here, the closest effort of at least mapping some more resilient cryptography for the post-quantum era is currently led by NIST, expected to mature circa 2022-2024 [128]. SAE is advised to already start preparing for the upcoming quantum computing era by forming some dedicated forum for this.

## Recommendations

Finally, this domain, of methodological uncertainties embedded in the DO-326/ED-202 set, may yet be the most challenging to settle, as it may require some unorthodox procedures, revisiting some of the most basic assumptions of the set. However, as the entire aviation safety-critical certifying community should be used for decades, assumptions should be validated and verified—and in this case—even the regulatory assumptions themselves might need to be validated and verified. The methodological assumptions reviewed here are:

**TABLE 2.** Safety Versus Security Minimum Requirements per Severity Levels.

| Severity Level | Safety DAL: Per *Element* | Security Measure(s) SAL: Per *Threat-Scenario* |
|---|---|---|
| Catastrophic | A | One SAL-2 + One SAL-3 |
| Hazardous | B | 3 |
| Major | C | 2 |
| Minor | D | 0 |
| No Safety Effect | E | 0 |

© AFUZION-InfoSec

- Should the DO-326/ED-202 set focus on processes alone, or does it need to also specify minimum performance requirements?

- Could the open-ended assurance process, relating security measures to scenarios rather than to systems/items even be viable?

- Is the focus on attacks ("Badness") such a good idea after all?

- Are we fooling ourselves by associating the likes of "Likelihood" or "Effectiveness" with the porous nature of cyber-threats (i.e., maybe, after all, we should treat any cyber-attack as imminent, just as we do for software defects)?

- Would actual risk assessments, when performed, eventually deem all systems "connected," so that we might not be able to treat them as isolated as far as safety? In other words, is this the end of federated avionics?

- To what extent would the current methodologies we apply be able to withstand future paradigm shifts such as AI, quantum computing, and maybe others that we are currently not even aware of?

As these uncertainties are by definition uncertain, they will probably need to be discussed first at the academic and professional levels. However, the relevant forums and committees should keep their eyes on the ball at all times.

# Summary/Conclusions

## Next Steps for Airworthiness Cybersecurity Regulation

This publication should be considered only as a first step toward clarifying the issues around Airworthiness Cybersecurity Regulation. The intention behind this and other SAE EDGE™ Research Reports is to start a dialogue among interested parties on important industry-wide topics that require further attention. The expectation is that these explorations of unsettled areas of technology will lead to the formation of working groups and, ultimately, committees that can address and resolve the

issues they raise, producing a framework for developing a common vocabulary of definitions, best practices, protocols, and standards needed to support continued progress toward safer and more innovative aviation and automotive products.

The experts' communications that gave rise to this publication demonstrated a great willingness throughout the aviation regulatory ecosystem to better define the terminology, procedures, and eventually the standards needed to enable the Airworthiness Cybersecurity Regulation to move ahead as quickly and efficiently as possible. SAE International has demonstrated its lead in this and closely related areas by the many experts contributing to this publication through a variety of channels, including, for instance, SAE authors, standard makers, and leaders on many disciplines.

This SAE EDGE™ Research Report on "Unsettled Topics Concerning Airworthiness Cybersecurity Regulation" identifies the following key topics for further pursuit, both through continued informal discussions among industry practitioners and through more formal working groups:

- Solidify legislation and regulation throughout the entire aviation ecosystem and beyond, in order to support the airworthiness cybersecurity complex certification, which is meant to ensure the safety of commercial passenger aircraft.

- Complete and enhance the aircraft-production supply chain cybersecurity regulation, in order to keep their current efficiency while ensuring their secure deliverables; devise proper strategies for keeping such supply chains from falling apart.

- Bring regulators, industry, and standard makers, from as many nations and organizations as practical together in order to fine-tune a proper balance, which would enable certification for most of the current airworthiness actors, while keeping the entire ecosystem at an acceptable level of safety and security.

- Standard makers of all relevant organizations to initiate discussions in order to fill major gaps in the current regulatory documents, settle dilemmas in these documents, and perform table-top reviews for methodology uncertainties.

# SAE EDGE™ Research Reports

SAE EDGE™ Research Reports, like the present report on "Unsettled Topics Concerning Airworthiness Cybersecurity Regulation," are intended to push further out into still unsettled areas of technology of interest to the mobility industry. SAE launches these reports before attempting to form a joint working group, let alone a cooperative research program or a standards committee.

SAE EDGE™ reports are intended to be quick, concise overviews of major unsettled areas where vital new technologies are emerging. An unsettled area is characterized more by confusion and controversy than established order. Early practitioners must confront an absence of agreement.

Their challenge is often not to seize the high ground but to find common ground. These scouting reports from the frontiers of investigation are intended merely to begin the process of sorting through critical issues, contributing to a better understanding of key problems, and providing helpful suggestions about possible next steps and avenues of investigation.

SAE EDGE™ Research Reports, therefore, are fundamentally distinct from the more formal working groups approach and far removed from the more mature research program and standard's development process.

# Recommendations

The overall recommendations of this SAE EDGE™ Research Report can be summarized as follows:

1. Aviation ecosystem and beyond

   - Cultivate cybersecurity regulation for aviation sectors beyond airworthiness, such as ATM, ANS, airports, and more.

   - Clearly define cybersecurity regulatory borders among such sectors.

   - Clearly define regulatory interfaces concerning cybersecurity across sector borderlines.

   - Formalize cybersecurity cooperation requirements among aviation stakeholders.

   - Initiate closer cybersecurity regulatory cooperation with ecosystems other than aviation.

   - Press on to form an *effective* international widely accepted trust framework.

2. OEMs supply chain

   - Better align cybersecurity regulatory requirements from lower-tier actors to fit the realities of such actors, even at the cost of increasing regulatory burden for OEMs and higher-tier actors.

   - Elevate the regulation for hardware cybersecurity beyond just counterfeit treatment to become equivalent (as practical) to software's.

   - Initiate cybersecurity table-top reviews of COTS and legacy systems and items to derive practical and regulatory consequences.

   - Make cybersecurity regulatory requirements for documents a two-way street, with requirements for OEM and high-tier suppliers' documentation to be provided to lower-tier actors.

   - Educate lower-tier actors with the new cybersecurity mandates.

   - Considerably enhance the regulation quality of internal development and production processes.

3. Regulatory compliance

   - Raise awareness of the benefits of proper cybersecurity housekeeping to counter resentment at all levels of applicants.

   - Initiate regulator-to-regulator cybersecurity coordination to mitigate regulatory clashes.

   - Initiate education programs for first time airworthiness applicants to mitigate "Rookie Wall" effects.

   - Mature applicants should be advised to start filling any ARP4754/4761 gaps they may still have, so as not to be caught by surprise when the new cybersecurity regulation hits them.

   - Carefully adjust the detail level, preferably in three tiers: legislation/regulation, guidance/recommendations, and technical/practical.

   - Carefully work toward a proper regulator-industry expertise-level balance.

4. Current standard gaps—standard makers are advised to enhance standards for any upcoming versions, to improve regulatory coverage of

   - Insider threats

   - Human factors in general

   - Non-technological issues and solutions in general

   - Resilience enhancement recommendations

   - Mandates for periodic detective/preventive activities

   - Practices for recertification after being compromised

5. Inherent dilemmas—standard makers are advised to reconcile inherent standard dilemmas for any upcoming versions, to minimize uncertainties

   - How simple or complex should the regulation be

   - The issue of an eternally evolving cybersecurity regulation

   - The black/white "trustworthiness" attribute conflict with the "aircraft domains" approach

   - The lacking proactive techniques in the DO-326/ED-202 set

   - The conflicting "slow" safety and "fast" security

   - Means to properly select a risk assessment methodology

6. Methodology uncertainties-standard makers are advised to initiate discussions, as a self-improvement process regarding

- Processes versus minimum performance requirements for the DO-326/ED-202 set

- Limitations and potential alleviation of the open-ended assurance process

- Reassessing the conceptual emphasis on "badness"

- What is the real value of the "Likelihood"/"Effectiveness" concept—and should it change?

- Would the federated avionics approach be deemed obsolete by cybersecurity considerations—and, if so, what to do about it?

- How to properly approach future paradigm shifts such as AI, quantum computing, and maybe others that we are currently not even aware of?

# Abbreviations/Definitions

**A-ISAC** - Aviation ISAC (Information Sharing and Analysis Center)

**AC** - Advisory Circular

**ACARS** - Aircraft Communications Addressing and Reporting System

**ADS-B** - Automatic Dependent Surveillance-Broadcast

**AI** - Artificial Intelligence

**AIA** - Aerospace Industries Association (of America, Inc.)

**AIAA** - American Institute of Aeronautics and Astronautics

**AISS** - Aeronautical Information System Security

**AMC** - Acceptable Means of Compliance

**ANN** - Artificial Neural Networks

**ANS** - Air Navigation Service(s)

**ATM** - Air Traffic Management

**BCG** - Boston Consulting Group

**BIT** - Built-in Test

**C2M2/CCMM** - Cybersecurity Capability Maturity Model

**CAA** - Civil Aviation Authority

**CAPEC** - Common Attack Pattern Enumeration and Classification

**C-BIT** - Continuous Built-in Test

**CC** - Common Criteria (ISO/IEC 15408)

**CIP** - Critical Infrastructure Protection

**CFTC** - Commodity Futures Trading Commission

**CMMC** - Capability Maturity Model for Cybersecurity

**CMMI** - Capability Maturity Model Integration

**CNCI** - Comprehensive National Cybersecurity Initiative

**COTS** - Commercial Off-the-Shelf

**CPS** - Cyber-Physical System(s)

**CPU** - Central Processing Unit

**CRAM** - Cyber Risk Assessment Model

**CRL** - Cyber Resilience Level

**CVE** - Common Vulnerabilities and Exposures

**CVSS** - Common Vulnerability Scoring System

**CWE** - Common Weakness Enumeration

**DAL** - Design Assurance Level

**DER** - Designated Engineering Representative

**DHS** - Department of Homeland Security

**DOA** - Design Organizations Approvals

**DOD** - Department of Defense

**EAL** - Evaluation Assurance Level

**EASA** - European Aviation Safety Agency

**ECCSA** - European Centre for Cybersecurity in Aviation

**EFB** - Electronic Flight Bag

**ENISA** - European Union Agency for Network and Information Security

**EUROCAE** - European Organisation for Civil Aviation Equipment

**FAA** - Federal Aviation Administration

**GDPR** - General Data Protection Regulation

**I-BIT** - Initial Built-in Test

**IATA** - International Air Transport Association

**IATF** - International Automotive Task Force

**ICAO** - International Civil Aviation Organization

**IEEE** - Institute of Electrical and Electronics Engineers

**IFEC** - In-Flight Entertainment and Communication

**IMA** - Integrated Modular Avionics

**ISAC** - Information Sharing and Analysis Center

**ISEM** - Information Security Event Management

**MRO** - Maintenance, Repair, and Overhaul

**NDA** - Nondisclosure Agreement

**NERC** - North American Electric Reliability Corporation

**NextGen** - Next-Generation Air Transportation System

**NICE** - National Initiative for Cybersecurity Education

**NIST** - National Institute of Standards and Technology

**NPA** - Notice of Proposed Amendment

**NSA** - National Security Agency

**NVD** - National Vulnerability Database

**OEM** - Original Equipment Manufacturer

**OWASP** - Open Web Application Security Project

**PED** - Personal Electronic Device

**RMT** - Rule-Making Task

**RPAS** - Remotely Piloted Aircraft System

**RTCA** - Radio Technical Commission for Aeronautics

**SAL** - Security Assurance Level

**SAMM** - Software Assurance Maturity Model

**SC** - Special Committee

**SCMH** - Supply Chain Management Handbook

**SESAR** - Single European Sky ATM Research

**SML** - Strength of Mechanism Level

**SOTA** - Software updates Over the Air

**SSGC** - Secretariat Study Group on Cybersecurity

**SSIG** - System Security Integration Guidance

**STPA-Sec** - System Theoretic Process Analysis Security

**TAME** - Trusted and Assured Micro-Electronics

**UAS** - Unmanned Aerial System

**UTM** - Urban Traffic Management

**WEF** - World Economic Forum

**WG** - Working Group

## Acknowledgments

Recognition should go first to all the participants, many of who also provided feedback on the draft version of this publication. Without their input and initiative, this SAE EDGE™ Research Report would not have been possible.

Dror Ben-David, *Neural Networks R&D Lab (NRDL) at Matrix*
Daniel DiMase, *Aerocyonics Inc.*
Vance Hilderman, *AFUZION-InfoSec*
Angeliki Karakoliou, *EASA*
Kirsten M. Koepsel, JD, *Independent Aviation Cybersecurity & Counterfeit Parts Expert*
Patrick Mana, *EUROCONTROL*

Daniel Nebenzahl, *Resilience Cyber Security*
Antonio Nogueras, *EUROCONTROL*
Chris Roberts, *Hillbilly Hit Squad*
Cyrille Rosay, *EASA*
Peter Skaves, *FAA Advisor*
Chris Sundberg, *Woodward, Inc.*

The author of this document together with the SAE Team responsible for its creation join in expressing their deepest appreciation to all the individuals mentioned above. It should be noted that these individuals, who contributed a wealth of indispensable information and views leading to this report, are by no means responsible for the views expressed in the report itself, nor do the views expressed here reflect in any way any specific views expressed by these individuals—the sole responsibility for these views lies with the author of this report.

Aharon David
AFUZION-InfoSec
Chief WHO (White Hat Officer)
(External consultant to SAE International)

## References

1. SITA, "FLYING DATA CENTERS TAKE OFF," Air-Transport IT-Review, Issue 1, 2015.

2. Daily Mail Online, "REVEALED: Security Expert Who 'Hacked a Commercial Flight and Made it Fly Sideways' Bragged that he also Hacked the International Space Station," 2015, https://www.dailymail.co.uk/news/article-3090288/Security-expert-admitted-FBI-took-control-commercial-flight-bragged-hacker-convention-2012-playing-International-Space-Station-getting-yelled-NASA.html, accessed January 29, 2020

3. Independent, "Ukraine says major cyberattack on Kiev's Boryspil airport was launched from Russia," https://www.independent.co.uk/news/world/europe/ukraine-cyberattack-boryspil-airport-kiev-russia-hack-a6818991.html, accessed January 29, 2020

4. Reuters, "Polish airline, hit by cyber attack, says all carriers are at risk," 2015, https://uk.reuters.com/article/us-poland-lot-cybercrime-idUKKBN0P21DC20150622, accessed January 29, 2020

5. FAA, 14 CFR Part 25, [Docket No. NM365 Special Conditions No. 25-357-SC], "Special Conditions: Boeing Model 787- 8 Airplane; Systems and Data Networks Security-Protection of Airplane Systems and Data Networks from Unauthorized External Access," 2007.

6. DoT FAA, "ARAC Report, Aircraft System Information Security/Protection," 2016, p. 20

7. FAA, "Policy Statement PS-AIR-21.16-02, "Establishment of Special Conditions for Aircraft Systems Information Security Protection," 2014.

8. FAA, "Policy Statement PS-AIR-21.16-02 Rev. 2, 'Establishment of Special Conditions for Aircraft Systems Information Security Protection,'" 2017.

9. EUROCAE WG-72 Aeronautical Systems Security, https://www.eurocae.net/about-us/working-groups/, accessed January 29, 2020

10. RTCA SC-216, "Aeronautical Systems Security," https://www.rtca.org/content/sc-216, accessed January 29, 2020

11. DoT FAA, "RAC Report, Aircraft System Information Security/Protection," 2016.

12. RTCA/EUROCAE, "DO-326A/ED-202A - Airworthiness Security Process Specification," 2014.

13. RTCA/EUROCAE, "DO-356A/ED-203A - Airworthiness Security Methods & Considerations," 2018.

14. RTCA/EUROCAE, "DO-355/ED-204 - Information Security Guidance for Continuing Airworthiness," 2014.

15. EUROCAE, "ED-201 - Aeronautical Information System Security (AISS) Framework Guidance," 2015.

16. EUROCAE, "ED-205 - Process Standard for Security Certification/Declaration of Air Traffic Management/Air Navigation Services (ATM/ANS) Ground Systems," 2019.

17. EUROCAE, "ER-013 - AERONAUTICAL INFORMATION SYSTEM SECURITY GLOSSARY," 2015.

18. EUROCAE, "ER-017 - INTERNATIONAL AERONAUTICAL INFORMATION SECURITY ACTIVITY MAPPING SUMMARY," 2018.

19. FAA, "Advisory Circular AC 119-1, Airworthiness & Operational Authorization of DO-326/ED-202 set FAA, Advisory Circular AC 20-140C, Guidelines for Design Approval of Aircraft Data Link Communication Systems Supporting Air Traffic Services (ATS)," 2015.

20. FAA, "Advisory Circular (AC) 20-140C (Sep 2016), Guidelines for Design Approval of Aircraft Data Link Communication Systems Supporting Air Traffic Services (ATS)," 2016.

21. FAA, "Advisory Circular AC 120-76D, Authorization for Use of Electronic Flight Bags", from 2017, includes a new "Security Procedures," 2017.

22. EASA RMT.0648, "Aircraft Cybersecurity," https://www.easa.europa.eu/document-library/rulemaking-subjects/aircraft-cybersecurity, accessed January 29, 2020

23. EASA, "Notice of Proposed Amendment 2019-01: Aircraft cybersecurity," 2019, https://www.easa.europa.eu/sites/default/files/dfu/NPA%202019-01.pdf, accessed January 29, 2020

24. EASA, "ED Decision 2020/010/R,"July 2019, https://www.easa.europa.eu/document-library/agency-decisions/ed-decision-2020010r, accessed Aug. 23, 2020.

25. UK CAA, "CAP 1753 - The Cyber Security Oversight Process for Aviation," 2019.

26. UK CAA, "CAP 1849 - Cyber Security Critical Systems Scoping Guidance," 2019.

27. UK CAA, "CAP 1850 - Cyber Assessment Framework (CAF) for Aviation Guidance," 2019.

28. UK DOT, "Aviation Cyber Security Strategy," 2018.

29. ICAO, ASSEMBLY - 40th SESSION, Resolution A40-10, 2019.

30. IATA, "AVIATION CYBER SECURITY - MOVING FORWARDS," ICAO 40th Assembly, 2019.

31. ISO/IEC 27000:2018, "Information technology - Security techniques - Information security management systems"

32. EASA RMT.0720, "Cybersecurity Risks," https://www.easa.europa.eu/sites/default/files/dfu/ToR%20RMT.0720.pdf, accessed January 29, 2020

33. EASA RMT.0648, "Aircraft Cybersecurity," https://www.easa.europa.eu/sites/default/files/dfu/ToR%20RMT.0648%20Issue%201.pdf, accessed January 29, 2020

34. Kovacs, E., "HITB2013AMS: Flaws in Aircraft Systems Allow Hackers to Hijack Airplanes," *Softpedia News*, 2013, https://news.softpedia.com/news/HITB2013AMS-Flaws-in-Aircraft-Systems-Allow-Hackers-to-Hijack-Airplanes-344754.shtml, accessed January 29, 2020

35. NIST Special Publication 800-82 Revision 2, "Guide to Industrial Control Systems (ICS) Security," 2015.

36. NIST, "Framework for Improving Critical Infrastructure Cybersecurity," Version 1.1, 2018.

37. ENISA, "Definition of Cybersecurity - Gaps and overlaps in standardization," 2015.

38. FinSAC, *Financial Sector's Cybersecurity: A Regulatory Digest* (World Bank Group, 2019).

39. Financial Stability Institute, "Regulatory approaches to enhance banks' cyber-security frameworks," 2017.

40. Boston Consulting Group, "RADICALLY SIMPLIFYING REGULATORY COMPLIANCE IN CYBERSECURITY," 2019.

41. CAO, ASSEMBLY - 40th SESSION, Resolution A40-10 (pp. 43-45), 2019.

42. ICAO, Annual Report 2018, Cybersecurity and Secretariat Study Group on Cybersecurity (SSGC), 2018.

43. ICAO, ASSEMBLY - 39th SESSION, Resolution A39-19 (pp. 91-93), 2016.

44. Levin, A., "Russian Hackers Attacked U.S. Aviation as Part of Breaches," Bloomberg, 2018, https://www.bloomberg.com/news/articles/2018-03-16/russian-hackers-penetrated-u-s-aviation-sector-early-last-year, accessed January 29, 2020

45. Doffman, Z., "Chinese Hackers Suspected Of Airbus Cyberattacks-A350 Among Targets," *Forbes*, 2019, https://www.forbes.com/sites/

zakdoffman/2019/09/26/china-suspected-of-multiple-airbus-cyberattacksa350-among-targets/?wpisrc=nl_cybersecurity202&wpmm=1#325aac99e630, accessed January 29, 2020

46. ICAO, ASSEMBLY - 40th SESSION, Report of the Technical Commission on Agenda Item 30 (p. 30-13), 2019.

47. AIA, "Civil Aviation Cybersecurity Industry Assessment & Recommendations," 2019, p. 26

48. NSA, Information Assurance Technical Framework (IATF) - Release 3.1, 2002, p. 4-3

49. WEF, Cyber Resilience Playbook for Public-Private Collaboration, 2017.

50. Lydon, B., "Industrial Cybersecurity & International Defense - Inside Siemens' Cybersecurity Charter of Trust," Automation.com, 2018, https://www.automation.com/automation-news/article/industrial-cybersecurity-international-defense-inside-siemens-cybersecurity-charter-of-trust, accessed January 29, 2020

51. "Around 80% of Airbus' activity is sourced. The company works with more than 12,000 suppliers worldwide that provide products and services for flying and non-flying parts," https://www.airbus.com/be-an-airbus-supplier.html, accessed January 29, 2020

52. "The Boeing supplier network includes more than 20,000 suppliers and partners," https://www.boeing.com/company/key-orgs/boeing-international/, accessed January 29, 2020

53. CheckPoint, Annual Cyber-Security Report, 2020, p. 24

54. Wainscott-Sargent, A., *Industry Responds to New Security Regulations, Vulnerabilities Facing Embedded Suppliers*, Avionics International, 2020, http://interactive.aviationtoday.com/avionicsmagazine/december-2019-january-2020/industry-responds-to-new-security-regulations-vulnerabilities-facing-embedded-suppliers/, accessed January 29, 2020

55. ISO/IEC 27036-1:2014, "Information technology - Security techniques - Information security for supplier relationships," 2014

56. NISTIR 7622, "Notional Supply Chain Risk Management Practices for Federal Information Systems," 2012.

57. NIST, Cyber Supply Chain Risk Management, https://csrc.nist.gov/Projects/cyber-supply-chain-risk-management, accessed January 29, 2020

58. U.S. 116TH CONGRESS, NATIONAL DEFENSE AUTHORIZATION ACT FOR FISCAL YEAR 2020, sections 224, 845

59. DFARS 252.204-7012, "Safeguarding Covered Defense Information and Cyber Incident Reporting," 2019.

60. Elovici, Y. et al., "dr0wncd-AM Cyber Attack," University of South Alabama, 2016, https://www.southalabama.edu/colleges/soc/computerscience/news/dr0wned.html, accessed January 29, 2020

61. Koepsel, K.M., *Counterfeit Parts and Their Impact on the Supply Chain*, 2nd Edition, SAE T-136, 2019.

62. Cimpanu, C., "Unpatchable security flaw found in popular SoC boards," ZDNet, 2019, https://www.zdnet.com/article/unpatchable-security-flaw-found-in-popular-soc-boards/, accessed January 29, 2020

63. SAE AS5553C, "Counterfeit Electronic Parts; Avoidance, Detection, Mitigation, and Disposition," 2019.

64. Trusted and Assured MicroElectronics Forum, https://tameforum.org/, accessed January 29, 2020

65. Collier, Z.A. et al, "A Semi-Quantitative Risk Assessment Standard for Counterfeit Electronics Detection," SAE, 2014.

66. DiMase, D. et al., *Systems Engineering Framework for Cyber Physical Security and Resilience* (Springer, 2015).

67. Robertson, J. and Riley, M., *The Big Hack: How China Used a Tiny Chip to Infiltrate U.S. Companies*, Bloomberg, 2018, https://www.bloomberg.com/news/features/2018-10-04/the-big-hack-how-china-used-a-tiny-chip-to-infiltrate-america-s-top-companies, accessed January 29, 2020

68. Porup, J.M., "Boeing's poor information security posture threatens passenger safety, national security, researcher says," CSO-online, 2019, https://www.csoonline.com/article/3451585/boeings-poor-information-security-posture-threatens-passenger-safety-national-security-researcher-s.html, accessed January 29, 2020

69. AIA, "Civil Aviation Cybersecurity Industry Assessment & Recommendations," 2019, p. 10

70. RTCA/EUROCAE, "DO-330/ED-215 - Software Tool Qualification Considerations," 2011.

71. Taleb, N., *The Black Swan: The Impact of the Highly Improbable* (Random House, 2007).

72. Cyber Insurance (AIG), https://www.aig.com/business/insurance/cyber-insurance, accessed January 29, 2020

73. Tine, D.R., "Microgrid Systems Design, Optimization, and Risk Drivers," Munich Re, 2017.

74. Travellers Insurance Co, "Cyber Risks for Solar and Wind Installations," https://www.travelers.com/business-insights/industries/energy/cyber-risks-for-solar-and-wind-installations, accessed January 29, 2020

75. NERC-CIP Compliance & Enforcement, https://www.nerc.com/pa/comp/Pages/default.aspx, accessed January 29, 2020

76. NERC-CIP Standards, https://www.nerc.com/pa/Stand/Pages/CIPStandards.aspx, accessed January 29, 2020

77. Montalbano, E., "Updated: Secrecy Reigns as NERC Fines Utilities $10M citing Serious Cyber Risks," *The Security Ledger*, 2020, https://securityledger.com/2019/02/secrecy-reigns-as-nerc-fines-utilities-10m-citing-serious-cyber-risks/, accessed January 29, 2020

78. IBM Security, "Cost of a Data Breach Report," 2019.

79. REGULATION (EU) 2016/679 OF THE EUROPEAN PARLIAMENT AND OF THE COUNCIL [General Data Protection Regulation (GDPR)], 2016.

80. AIA, "Civil Aviation Cybersecurity Industry Assessment & Recommendations," 2019, p. 13.

81. SAE, "ARP4754A: Guidelines for Development of Civil Aircraft and Systems," 2010.

82. SAE, "ARP4761 - Guidelines and Methods for Conducting the Safety Assessment Process on Civil Airborne Systems and Equipment," 1996.

83. RTCA/EUROCAE, "DO-178C/ED-12C - Software Considerations in Airborne Systems and Equipment Certification," 2012.

84. RTCA/EUROCAE, "DO-254/ED-80, Design Assurance Guidance for Airborne Electronic Hardware," 2000.

85. Hilderman, V., "Introduction to DO-178C," SAE, https://www.sae.org/learn/content/c1410/, accessed January 29, 2020

86. Hilderman, V., "Avionics Software Advanced DO-178C Training Workshop," AFUZION, https://afuzion.com/private-training/avionics-software-advanced-do-178c-training-class/, accessed January 29, 2020

87. Hilderman, V., "Applying DO-254 for Avionics Hardware Development and Certification," SAE, https://www.sae.org/learn/content/c1703/, accessed January 29, 2020

88. Hilderman, V., "Avionics Hardware Intermediate DO-254 Training Workshop," AFUZION, https://afuzion.com/private-training/avionics-hardware-intermediate-do-254-training-class/, accessed January 29, 2020

89. Peterson, E.M., "ARP4754A and the Guidelines for Development of Civil Aircraft and Systems," SAE, https://www.sae.org/learn/content/c1118/, accessed January 29, 2020

90. Hilderman, V., "Avionics Systems ARP4754A Training Workshop," AFUZION, https://afuzion.com/private-training/avionics-systems-arp-4754a-training-class/, accessed January 29, 2020

91. Peterson, E.M., "ARP4761 and the Safety Assessment Process for Civil Airborne Systems," SAE, https://www.sae.org/learn/content/c1245/, accessed January 29, 2020

92. Hilderman, V., "ARP4761A Training: Aircraft & System Safety Training," AFUZION, https://afuzion.com/private-training/arp-4761a-training-aviation-safety-training/, accessed January 29, 2020

93. David, A., "DO -326A and ED-202A : An Introduction to the New and Mandatory Aviation Cyber-Security Essentials," SAE, https://www.sae.org/learn/content/c1949/, accessed January 29, 2020

94. David, A., "DO-326A / ED-202A Training: Aviation Cyber Security," AFUZION, https://afuzion.com/private-training/do-326a-ed-202a-training-aviation-cyber-security/, accessed January 29, 2020

95. Henry, J., "These 5 Types of Insider Threats Could Lead to Costly Data Breaches," *Security Intelligence*, 2018, https://securityintelligence.com/these-5-types-of-insider-threats-could-lead-to-costly-data-breaches/, accessed January 29, 2020

96. EC-Council, "5 EMERGING CYBER THREATS TO WATCH IN 2020," 2019; https://blog.eccouncil.org/5-emerging-cyber-threats-to-watch-in-2020/, accessed January 29, 2020

97. The Cyber Kill Chain', https://www.lockheedmartin.com/en-us/capabilities/cyber/cyber-kill-chain.html, accessed January 29, 2020

98. Cooper, Pete, "AVIATION CYBERSECURITY-FINDING LIFT, MINIMIZING DRAG," p. 53, Atlantic Council - Scowcroft Center for Strategy and Security, 2017.

99. NSA, "Information Assurance Technical Framework (IATF) - Release 3.1," 2002; p. 4-40.

100. CFTC, "17 CFR Parts 37, 38, and 49: System Safeguards Testing Requirements; Final Rule," 2016.

101. AIA, "Civil Aviation Cybersecurity Industry Assessment & Recommendations," 2019, p. 18

102. DHS, "Cybersecurity Capability Maturity Model White Paper," 2014.

103. DoD, "Cybersecurity Maturity Model Certification," 2020.

104. Cyber Resiliency Level™ (CRL™), https://lockheedmartin.com/en-us/capabilities/cyber/cyber-resiliency-level.html, accessed January 29, 2020

105. CMU/SEI, "CMMI® for Development, Version 1.3," 2010.

106. AIA, "Civil Aviation Cybersecurity Industry Assessment & Recommendations," 2019, p. 9

107. ARINC, "ARINC 664P5 - AIRCRAFT DATA NETWORK PART 5 NETWORK DOMAIN CHARACTERISTICS AND INTERCONNECTION," 2005.

108. ARINC, "ARINC REPORT 811 - COMMERCIAL AIRCRAFT INFORMATION SECURITY CONCEPTS OF OPERATION AND PROCESS FRAMEWORK," 2005.

109. DHS, "Moving Target Defense," 2011, https://www.dhs.gov/science-and-technology/csd-mtd, accessed January 29, 2020

110. Lincoln Laboratory, "Survey of Cyber Moving Targets," 2012, https://apps.dtic.mil/dtic/tr/fulltext/u2/a591804.pdf, accessed January 29, 2020

111. Johnson, J., *Implementing Active Defense Systems on Private Networks*, SANS Institute, 2020, https://www.sans.org/reading-room/whitepapers/detection/implementing-active-defense-systems-private-networks-34312, accessed January 29, 2020

112. DoT FAA, "ARAC Report, Aircraft System Information Security/Protection," 2016, p. 8, Recommendation 06

113. Safeware Engineering, "Safety-Critical Requirements Specification and Analysis using SpecTRM," 2003; "… Almost all accidents related to software components in the past 20 years can be traced to flaws in the requirements specifications …," http://www.safeware-eng.com/system%20and%20software%20safety%20publications/Requirements%20Specification.htm ; accessed January 29, 2020

114. NSA, Information Assurance Technical Framework (IATF) - Release 3.1, 2002; p. 4-32

115. Common Criteria for Information Technology Security Evaluation a.k.a., ISO/IEC 15408 - Release 3.1, 2012.

116. NSA, Information Assurance Technical Framework (IATF) - Release 3.1, 2002,p. 4-34

117. Wainscott-Sargent, A., *Can Blockchain Enhance Aviation Data Security?*, Avionics International, 2020, https://www.aviationtoday.com/2020/01/28/can-blockchain-enhance-aviation-data-security/, accessed January 29, 2020

118. NIST, National Vulnerability Database, https://nvd.nist.gov/vuln/full-listing, accessed January 29, 2020

119. Ranum, Marcus, "The Six Dumbest Ideas in Computer Security," 2005, http://www.ranum.com/security/computer_security/editorials/dumb/, accessed January 29, 2020

120. NIST, NVD CWE Slice, https://nvd.nist.gov/vuln/categories, accessed January 29, 2020

121. NIST, CVSS 3.1 Official Support, https://nvd.nist.gov/General/News/CVSS-v3-1-Official-Support, accessed January 29, 2020

122. NIST, Common Attack Pattern Enumeration and Classification (CAPEC), https://samate.nist.gov/BF/Enlightenment/CAPEC.html, accessed January 29, 2020

123. Butler, J.A., "Application and Evaluation of Built-In-Test (BIT) Techniques in Building Safe Systems," *CROSSTALK - The Journal of Defense Software Engineering*, 2006.

124. Nasser, A.M.K. et al, "Accelerated Secure Boot for Real-Time Embedded Safety Systems," SAE 11-02-01-0003, 2019.

125. IEC 62304:2006/AMD 1:2015, "Medical device software - Software life cycle processes - Amendment 1," 2015.

126. FAA-AC/EASA-AMJ 25.1309-1 (B-Arsenal Draft), "System Design and Analysis," section 5.k, 2002.

127. RTCA/EUROCAE, "DO-297/ED-124 - Integrated Modular Avionics (IMA) Development Guidance and Certification Considerations," 2005.

128. NISTIR 8240, "Status Report on the First Round of the NIST Post-Quantum Cryptography Standardization Process," NIST, 2019.

# Contact Information

EDGEresearch@sae.org.